# HARCOURT
# Science

## Harcourt School Publ

Orlando • Boston • Dallas • Chicago

www.harcourtschool.com

D1405048

**Cover Image:** This reptile is a veiled chameleon *(Chamaelo calyptratus).* They are often raised and sold as pets. In the wild, they live in trees on the humid parts of the Arabian peninsula.

Printed in the United States of America

ISBN 0-15-315692-9    UNIT A
ISBN 0-15-315693-7    UNIT B
ISBN 0-15-315694-5    UNIT C
ISBN 0-15-315695-3    UNIT D
ISBN 0-15-315696-1    UNIT E
ISBN 0-15-315697-X    UNIT F

2 3 4 5 6 7 8 9 10    032    2002  2001  2000

## Authors

**Marjorie Slavick Frank**
Former Adjunct Faculty Member at Hunter, Brooklyn, and Manhattan Colleges
New York, New York

**Robert M. Jones**
Professor of Education
University of Houston-Clear Lake
Houston, Texas

**Gerald H. Krockover**
Professor of Earth and Atmospheric Science Education
School Mathematics and Science Center
Purdue University
West Lafayette, Indiana

**Mozell P. Lang**
Science Education Consultant
Michigan Department of Education
Lansing, Michigan

**Joyce C. McLeod**
Visiting Professor
Rollins College
Winter Park, Florida

**Carol J. Valenta**
Vice President—Education, Exhibits, and Programs
St. Louis Science Center
St. Louis, Missouri

**Barry A. Van Deman**
Science Program Director
Arlington, Virginia

**UNIT A**

LIFE SCIENCE

# A World of Living Things

# UNIT B

## LIFE SCIENCE
# Looking at Ecosystems

# UNIT C

### EARTH SCIENCE
# Earth's Surface

# UNIT D

## EARTH SCIENCE
# Patterns on Earth and In Space

## UNIT E

### PHYSICAL SCIENCE

# Matter and Energy

**UNIT F**

PHYSICAL SCIENCE

# Forces and Motion

# Using Science Process Skills

When scientists try to find an answer to a question or do an experiment, they use thinking tools called the process skills. You use many of the process skills whenever you speak, listen, read, write, or think. Think about how these students used process skills to help them answer questions and do experiments.

**M**atthew is spending the day at the beach. He finds seashells. He carefully **observes** the shells and **compares** their shapes and their colors. He **classifies** them into groups according to their shapes.

**Try This** Observe, compare, and classify objects that interest you, such as rocks or leaves.

**Talk About It** How did Matthew use the skills of observing and comparing to classify his shells into groups?

### Process Skills

**Observe**—use the senses to learn about objects and events

**Compare**—identify characteristics of things or events to find out how they are alike and different

**Classify**—group or organize objects or events in categories based on specific characteristics

**L**ing wanted to find out whether sand rubbing against rocks would cause pieces of the rock to flake off. He collected three rocks, measured their masses, and then put them in a jar with sand and water. He shook the rocks every day for a week. At the end of the week he **measured** and **recorded** the mass of the rocks and the mass of the sand and the container. He **interpreted** his data and **concluded** that rocks are broken down when sand rubs against them.

**Try This** Use a thermometer to measure the temperature inside and outside your classroom at the same time each day for a week. Record, display, and interpret your data to find the average indoor and outdoor temperatures for the week.

**Talk About It** How does displaying your data in charts, tables, and graphs help you interpret it?

## Process Skills

**Measure** — compare an attribute of an object, such as mass, length, or capacity, to a unit of measure such as gram, centimeter, or liter.

**Gather, Record, Display, or Interpret Data**

- gather data by making observations which are used to make inferences or predictions
- record data by writing down the observations
- display data by making tables, charts, or graphs
- interpret data by drawing conclusions about what the data shows

**C**aitlin wanted to know how the light switch in her bedroom worked. She decided to **use a model** to see how the electrical wires in the wall and the switch worked to turn the light on and off. She used batteries, wires, a flashlight bulb, a bulb holder, thumbtacks, and a paper clip to build her model. She **predicted** that the bulb, the wires, and the batteries had to be connected to make the bulb light. She **inferred** that the paper clip switch interrupted the flow of electricity to turn off the light. Caitlin's model verified her prediction and her inference.

**Try This** Make a model to show how you can light more than one bulb.

**Talk About It** How does using a model help you understand how electricity works to light a bulb?

## Process Skills

**Use a model —** make a representation to explain an idea, an object, or an event, such as how something works

**Predict —** form an idea of an expected outcome based on observations or experience

**Infer —** use logical reasoning to explain events and make conclusions based on observations

**K**endra wants to know which brand of paper towel absorbs the most water. She **planned and conducted a simple investigation** to find out. She chose three brands of paper towels. She poured one liter of water into each of three beakers. She put a towel from each of the three brands in a beaker for 10 seconds. She pulled the towel out of the water and let it drain back into the beaker for 5 seconds. She then measured the amount of water left in the beaker.

She **controlled variables** in her experiment by making sure each beaker contained exactly the same amount of water and that she timed each step in her experiment exactly. Based on the results of this test, she was able to tell her Dad which brand of paper towel was the most absorbent.

**Try This** Plan and conduct an investigation to compare different brands of a product or service that you and your family use. Identify the variables that you will control.

**Talk About It** Why is it important to identify and control the variables in an investigation?

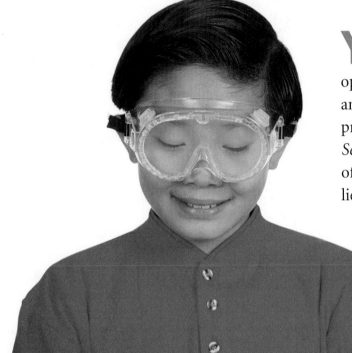

You will have many opportunities to practice and apply these and other process skills in *Harcourt Science.* An exciting year of science discoveries lies ahead!

# Safety in Science

Doing investigations in science can be fun, but you need to be sure you do them safely. Here are some rules to follow.

**1 Think ahead.** Study the steps of the investigation so you know what to expect. If you have any questions, ask your teacher. Be sure you understand any safety symbols that are shown.

**2 Be neat.** Keep your work area clean. If you have long hair, pull it back so it doesn't get in the way. Roll or push up long sleeves to keep them away from your experiment.

**3 Oops!** If you should spill or break something, or get cut, tell your teacher right away.

**4 Watch your eyes.** Wear safety goggles anytime you are directed to do so. If you get anything in your eyes, tell your teacher right away.

**5 Yuck!** Never eat or drink anything during a science activity unless you are told to do so by your teacher.

**6 Don't get shocked.** Be especially careful if an electric appliance is used. Be sure that electric cords are in a safe place where you can't trip over them. Don't ever pull a plug out of an outlet by pulling on the cord.

**7 Keep it clean.** Always clean up when you have finished. Put everything away and wipe your work area. Wash your hands.

In some activities you will see these symbols. They are signs for what you need to act safely.

**CAUTION**
Be especially careful.

**CAUTION**
Wear safety goggles.

**CAUTION**
Be careful with sharp objects.

**CAUTION**
Don't get burned.

**CAUTION**
Protect your clothes.

**CAUTION**
Protect your hands with mitts.

**CAUTION**
Be careful with electricity.

# Patterns on Earth and in Space

**Unit Project**

## Weather Station

Make a weather station for your classroom. Make instruments to help you measure the weather. Use charts and tables to record changes in the weather. Prepare a weather report each day, and predict the weather for the following day. Be sure to support your prediction with data you've collected. Then compare your prediction with the actual weather.

# Weather Conditions

E veryone talks about the weather, but weather forecasters get paid to talk about it. Many people depend on weather forecasts to plan their day. Sometimes forecasts of severe weather can even save lives.

### Vocabulary Preview

atmosphere
air pressure
troposphere
stratosphere
greenhouse effect
air mass
front
barometer
humidity
anemometer

### FAST FACT

Right now, 2000 thunderstorms are happening around the Earth. While you are reading this sentence, lightning will strike the Earth about 500 times!

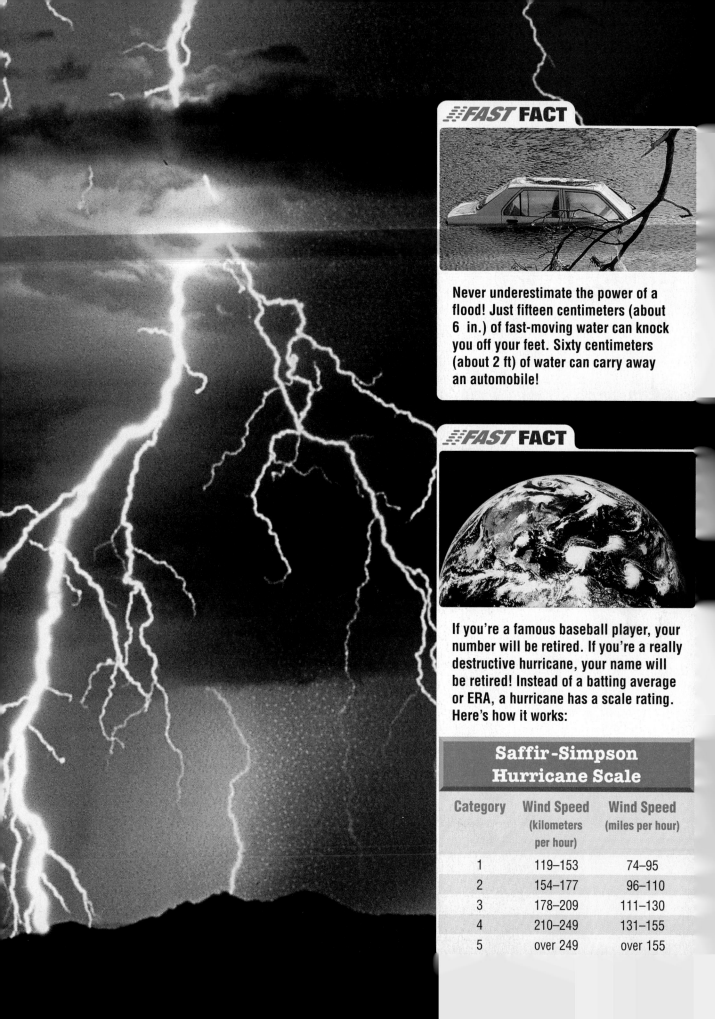

Never underestimate the power of a flood! Just fifteen centimeters (about 6 in.) of fast-moving water can knock you off your feet. Sixty centimeters (about 2 ft) of water can carry away an automobile!

If you're a famous baseball player, your number will be retired. If you're a really destructive hurricane, your name will be retired! Instead of a batting average or ERA, a hurricane has a scale rating. Here's how it works:

## Saffir-Simpson Hurricane Scale

| Category | Wind Speed (kilometers per hour) | Wind Speed (miles per hour) |
|---|---|---|
| 1 | 119–153 | 74–95 |
| 2 | 154–177 | 96–110 |
| 3 | 178–209 | 111–130 |
| 4 | 210–249 | 131–155 |
| 5 | over 249 | over 155 |

# LESSON 1

## What Makes Up Earth's Atmosphere?

In this lesson, you can . . .

 **INVESTIGATE** a property of air.

 **LEARN ABOUT** Earth's atmosphere.

 **LINK** to math, writing, art, and technology.

# A Property of Air

**Activity Purpose** Everything around you is matter. Matter is anything that takes up space and has weight. In this investigation you will **observe** a property of air. Then you will **infer** whether air is matter.

## Materials

- metric ruler
- piece of string about 80 cm long
- scissors
- 2 round balloons (same size)
- safety goggles
- straight pin

CAUTION

## Activity Procedure

**1** Work with a partner. Use the scissors to carefully cut the string into three equal pieces.
**CAUTION** Be careful when using scissors.

**2** Tie one piece of the string to the middle of the ruler.

◄ Oxygen is part of the air you breathe. High on a mountain the particles of air are far apart. The climber can't get enough oxygen from the air. He needs extra oxygen from a tank to keep his body working properly.

**3** Blow up the balloons so they are about the same size. Seal the balloons. Then tie a piece of string around the neck of each balloon.

**4** Tie a balloon to each end of the ruler. Hold the middle string up so that the ruler hangs from it. Move the strings so that the ruler is balanced. (Picture A)

**5** **CAUTION** **Put on your safety goggles.** Use the straight pin to pop one of the balloons. **Observe** what happens to the ruler.

Picture A

## Draw Conclusions

1. Explain how this investigation shows that air takes up space.

2. Describe what happened when one balloon was popped. What property of air caused what you **observed?**

3. **Scientists at Work** Scientists often **infer** conclusions when the answer to a question is not clear or can't be **observed** directly. Your breath is invisible, but you observed how it made the balloons and the ruler behave. Even though you can't see air, what can you infer about whether or not air is matter? Explain.

**Investigate Further** The air around you presses on you and everything else on Earth. This property of air, called air pressure, is a result of air's weight. When more air is packed into a small space, air pressure increases. You can feel air pressure for yourself. Hold your hands around a partly filled balloon while your partner blows it up. Describe what happens. Then **infer** which property of air helps keep the tires of a car inflated.

> **Process Skill Tip**
>
> Observations and inferences are different things. An **observation** is made with your senses. An **inference** is an opinion based on what you have observed and what you know about a situation.

D5

# Earth's Atmosphere

## The Air You Breathe

**FIND OUT**

• some properties of air

• the layers of the atmosphere

**VOCABULARY**

atmosphere
air pressure
troposphere
stratosphere

You can live for a few days without water and for many days without food. But you can live only a few minutes without air. Nearly all living things need air to carry out their life processes. The layer of air that surrounds our planet is called the **atmosphere** (AT•muhs•feer). When compared to the size of Earth, the atmosphere looks like a very thin blanket surrounding the entire planet.

The atmosphere wasn't always as it is today. It formed millions of years ago as gases from erupting volcanoes collected around the planet. This mixture of gases would have poisoned you if you had breathed it. But bacteria and other living things used gases in this early atmosphere. They released new gases as they carried out their life processes. Over time, the gas mixture changed slowly to become the atmosphere Earth has now.

The atmosphere now is made up of billions and billions of gas particles. Almost four-fifths of these gas particles are nitrogen. Oxygen, a gas that your body uses in its life processes, makes up about one-fifth of the atmosphere. Other gases, including carbon dioxide and water vapor, make up the rest of the atmosphere.

Although you can't see all of it, a thin blanket of air called the atmosphere surrounds Earth. ▼

Plants use carbon dioxide during the process of photosynthesis. Plants give off oxygen as photosynthesis occurs. Carbon dioxide also absorbs heat energy from the sun and from Earth's surface. This helps keep the planet warm.

Like carbon dioxide, water vapor can absorb heat energy. The amount of water vapor in the air varies from place to place. Air over bodies of water usually contains more water vapor than air over land. High in the air, water vapor condenses to form clouds.

Air has certain properties. As you saw in the investigation, air takes up space and has weight. All the particles of air pressing down on the surface cause **air pressure** (PRESH•er). Air pressure changes as you go higher in the atmosphere. The picture shows what a column of air might look like. At the surface of Earth, air particles are close together. The higher you go in the atmosphere, the farther apart the air particles are. So the air pressure is less as you go higher in the atmosphere.

### ✔ What is the atmosphere?

**1** Air particles in the upper atmosphere have the least weight pressing on them. The particles are far apart. Air in this part of the atmosphere is much less dense than air lower in Earth's atmosphere.

**2** Air near the middle of the atmosphere has more weight pressing down on it. So it is denser than air higher above Earth.

**3** The weight of the entire column of air presses down on the air particles closest to Earth, forcing them close together. This makes air densest at Earth's surface. Air pressure is greatest where air is densest.

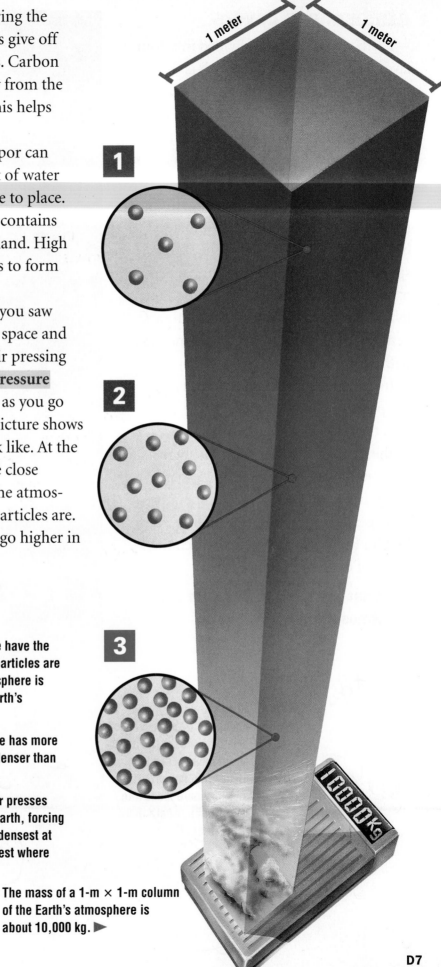

The mass of a 1-m × 1-m column of the Earth's atmosphere is about 10,000 kg. ▶

## Atmosphere Layers

Earth's atmosphere is divided into four layers. The layer closest to Earth is the **troposphere** (TROH•poh•sfeer). We live in the troposphere and breathe its air. Almost all weather happens in this layer. In the troposphere, air temperature decreases as you go higher.

Some airplanes that travel long distances fly in the **stratosphere** (STRAT•uh•sfeer) to be above most bad weather. The stratosphere contains most of the atmosphere's ozone, a kind of oxygen. The ozone protects living things from the sun's harmful rays. Temperatures in the stratosphere increase with height.

In the mesosphere (MES•oh•sfeer), air temperature decreases with height. In fact, the mesosphere is the coldest layer of the atmosphere. The thermosphere (THER•moh•sfeer) is the hot, outermost layer of air. In the thermosphere, temperature increases quickly with height. Temperatures high in the thermosphere can reach thousands of degrees Celsius.

✔ **What are the four layers of the atmosphere?**

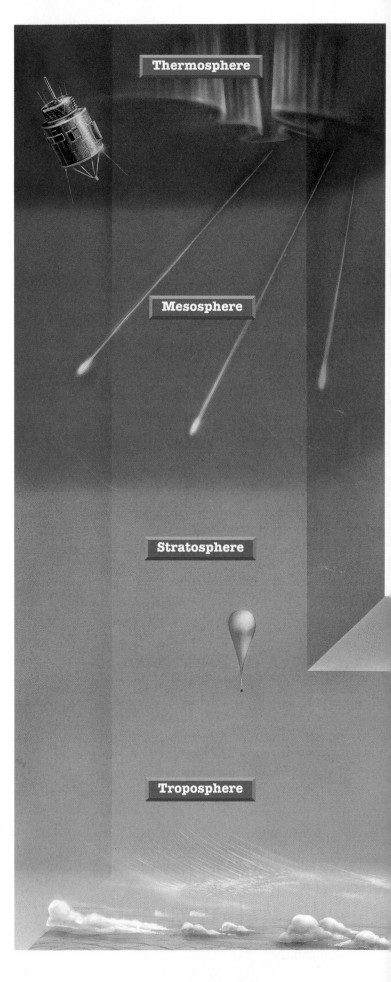

Earth's atmosphere is divided into four layers based on changes in air temperature. Each layer blends into the next. The thermosphere fades into outer space, where there is no air at all. ►

## Summary

The thin blanket of air that surrounds Earth is called the atmosphere. Earth's atmosphere is divided into four layers based on changes in temperature. The layers, starting with the one closest to Earth, are the troposphere, stratosphere, mesosphere, and thermosphere.

## Review

1. What is the atmosphere?
2. How does air pressure change with height?
3. How is the atmosphere divided?
4. **Critical Thinking** Compare and contrast the stratosphere and the mesosphere.
5. **Test Prep** In which layer of the atmosphere does most weather occur?
   A troposphere
   B stratosphere
   C mesosphere
   D thermosphere

# LINKS

## MATH LINK

**Atmospheric Temperatures** In the troposphere the air temperature drops about $6\frac{1}{2}$°C for every 1 kilometer increase in height. If the troposphere is about 10 kilometers thick and the air temperature at the ground is 30°C, what is the temperature at a height of 2 kilometers?

## WRITING LINK

**Informative Writing—Description** Pretend that you are falling from space toward Earth. For your teacher, write a story describing what you see and feel as you go through each layer of the atmosphere.

## ART LINK

**Atmosphere Layers** Paint a picture showing the atmosphere as you would see it from space. Label the layers.

## TECHNOLOGY LINK

Learn more about Earth's atmosphere and weather by visiting this Internet site.
**www.scilinks.org/harcourt**

SC**L**INKS™
THE WORLD'S A CLICK AWAY

# How Do Air Masses Affect Weather?

In this lesson, you can . . .

**INVESTIGATE** wind speed.

**LEARN ABOUT** what causes weather.

**LINK** to math, writing, health, and technology.

Wind, which is air in motion, keeps these kites fluttering in the sky. ▼

---

# INVESTIGATE

# Wind Speed

**Activity Purpose**  Have you ever flown a kite? A strong wind makes the kite flutter and soar through the air. A gentle breeze is usually not enough to keep the kite flying. What is wind? Wind is air in motion. In this investigation you will make an instrument to **measure** wind speed.

### Materials

- sheet of construction paper
- tape
- hole punch
- 4 gummed reinforcements
- glue
- piece of yarn about 20 cm long
- strips of tissue paper, about 1 cm wide and 20 cm long

---

## Activity Procedure

**1** Form a cylinder with the sheet of construction paper. Tape the edge of the paper to keep the cylinder from opening.

**2** Use the hole punch to make two holes at one end of the cylinder. Punch them on opposite sides of the cylinder and about 3 cm from the end. Put two gummed reinforcements on each hole, one on the inside and one on the outside. (Picture A)

**3** Thread the yarn through the holes, and tie it tightly to form a handle loop.

| Wind Scale | | | |
|---|---|---|---|
| **Speed (km/h)** | **Description** | **Objects Affected** | **Windsock Position** |
| 0 | no breeze | no movement of wind | |
| 6–19 | light breeze | leaves rustle, wind vanes move, wind felt on face | |
| 20–38 | moderate breeze | dust and paper blow, small branches sway | |
| 39–49 | strong breeze | umbrellas hard to open, large branches sway | |

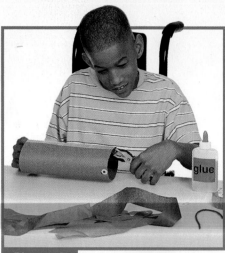

Picture A

**4** Glue strips of tissue paper to the other end of the cylinder. Put tape over the glued strips to hold them better. Your completed windsock should look like the one shown in Picture B.

**5** Hang your windsock outside. Use the chart above to **measure** wind speed each day for several days. **Record** your measurements in a chart. Include the date, time of day, observations of objects affected by the wind, and the approximate wind speed.

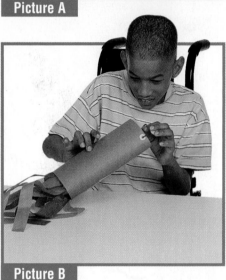

Picture B

## Draw Conclusions

1. How fast was the weakest wind you **measured**? How fast was the strongest wind?

2. How did you determine the speed of the wind?

3. **Scientists at Work** *Light*, *moderate*, and *strong* are adjectives describing wind speed. Scientists often use number **measurements** to describe things because, in science, numbers are more exact than words. What is the wind speed measurement in kilometers per hour if the wind is making large tree branches sway?

**Investigate Further** Use a magnetic compass to determine which way is north from your windsock. **Measure** both wind speed and direction each day for a week. **Record** your data in a chart.

# Air and Weather

## Air and the Sun

**FIND OUT**

- **how the sun affects weather**
- **what makes an air mass**

**VOCABULARY**

greenhouse effect
air mass
front

Have you ever watched a weather report on television? If so, you know that temperature, air pressure, and wind are some of the things reported. You also know that these weather conditions change every day. But do you know why?

Weather begins with the sun, which provides energy for making weather. But the amount of the sun's energy reaching Earth is not the same everywhere. More energy reaches the equator than the poles. This uneven heating is part of what causes air to move and what makes weather.

Most of the sun's energy never reaches Earth. It is lost in space. Of the tiny fraction of the sun's energy that does reach Earth, about three-tenths is reflected out into space. Another three-tenths warms the air. The other four-tenths warms the land and oceans. The atmosphere traps this heat much like the glass of a greenhouse. Without this **greenhouse effect**, Earth would reflect most of the sun's energy back into space and Earth's surface would be too cold to support life.

✔ **How does the atmosphere work like a greenhouse?**

▲ Have you ever seen ripples like these above a paved road on a hot day? The air just above the hot pavement is also hot. Light travels differently through hot air. That's why the view is blurry.

Sunlight passes through the atmosphere and warms Earth's surface. The greenhouse effect keeps most of the heat from escaping back into space. ▼

Not to scale

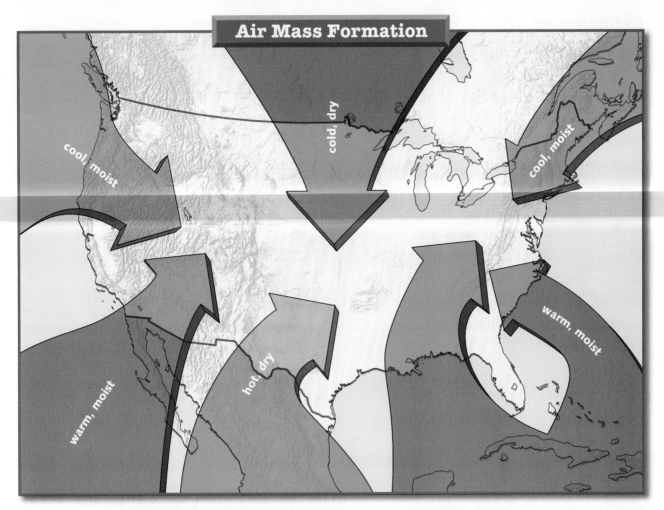

## Air Mass Formation

cool, moist

cold, dry

cool, moist

warm, moist

hot, dry

warm, moist

▲ Air masses form over both land and water. The map shows where the air masses that affect North America form. Cool air masses are in blue colors. Warm air masses are in red colors.

## Air Masses

If you could see the air around Earth from outer space, you would see large clumps of it forming, moving over Earth's surface, and slowly changing. These huge bodies of air, which can cover thousands of kilometers, are called air masses.

Like air heated by a hot road, an **air mass** has the same general properties as the land or water over which it forms. Two properties—moisture content and temperature—are used to describe air masses. Moist air masses form

over water. Air masses that form over land are generally dry. Air masses that form near Earth's poles are cold. Air masses that form in the tropics, or areas near the equator, are warm.

The map shows air masses forming and moving over the North American continent. You can see a polar air mass bringing cold, dry air from the north into the United States. You can also see warm, moist air coming in from the south as part of tropical air masses.

✓ **What is an air mass?**

## Air Masses Meet

Look again at the map on page D13. What do you think happens when different air masses meet? When two air masses meet, they usually don't mix. Instead they form a border called a **front**. Most of what you think of as weather happens along fronts.

A cold front is shown on the map just to the right. It forms when a cold air mass catches up to a warm air mass. The colder air mass forces the warmer air up into the atmosphere. As the warm air is pushed upward, it cools and forms clouds. Rain develops. Thunderstorms often occur along a cold front.

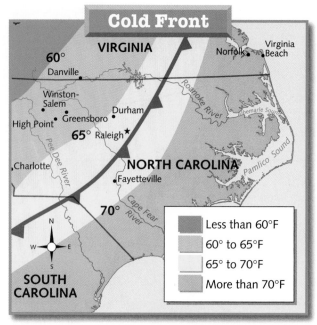

▲ A line with triangles is the symbol for a cold front. The air is colder behind a cold front than ahead of it. The triangles point in the direction of movement. In which direction is this front moving?

## THE INSIDE STORY

### Storm Front

When two air masses meet, they form a front. Thunderstorms and high winds often happen as a cold front moves through an area. After the front has passed, wind speed is lower, the sky is clear of clouds, and the temperature is lower.

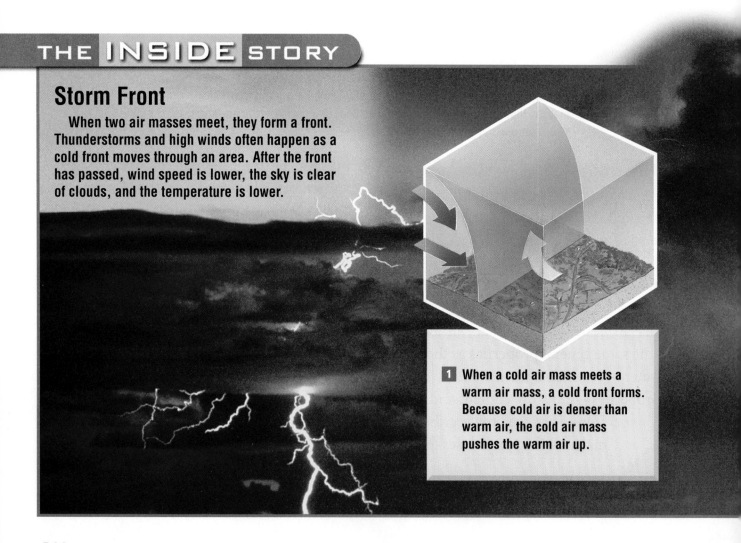

**1** When a cold air mass meets a warm air mass, a cold front forms. Because cold air is denser than warm air, the cold air mass pushes the warm air up.

**Warm Front**

VIRGINIA
Virginia Beach
Norfolk
Danville
60°
Winston-Salem
High Point
Greensboro •Durham
Raleigh★
65°
Roanoke River
Albemarle Sound
NORTH CAROLINA
Pamlico Sound
Charlotte
Fayetteville
70°
Pee Dee River
Cape Fear River
SOUTH CAROLINA

N
W E
S

Less than 60°F
60° to 65°F
65° to 70°F
More than 70°F

▲ A line with half-circles is the symbol for a warm front. The half-circles point in the direction the front is moving. The air is warmer behind this front than ahead of it.

A warm front forms when a warm air mass catches up to a cold air mass. The warm air slides up over the colder, denser air. Clouds form, sometimes many miles ahead of where the front is moving along the ground. Steady rain or snow may fall as the front approaches and passes. Then the sky becomes clear of clouds and the temperature becomes higher.

Sometimes a front stops moving. Such a front is called a stationary front. A stationary front can stay in one place for several days. The constant fall of snow or rain along a stationary front can leave behind many inches of snow or cause a flood.

✔ **What is a front?**

**2** As the warm air is forced up, it cools. It can no longer have as much water vapor. The extra water vapor begins to form clouds.

**3** Dense, puffy clouds with flat bottoms form along cold fronts. Sometimes these clouds are called thunderheads. They often produce lightning, thunder, and lots of rain in a short time.

## Air Masses Move

You can see air masses moving from place to place by watching how weather forms and changes. In the investigation you built a device to measure wind speed. Wind speed often increases as a front approaches. Wind direction also changes.

Air pressure also changes as air masses move over an area. As a front moves closer, air pressure drops. Air pressure rises as the front moves over the area.

Temperature, too, changes as a front moves over an area. Warmer air is brought into a region by a warm front. Likewise, the temperature goes down when a cold front moves over an area.

▲ The wind direction on each side of this cold front is shown by the arrows.

✔ **How does air pressure change as a front moves toward and then over an area?**

◄ The arrow of a weather vane points to the direction from which the wind is blowing.

▲ The same cold front has moved and changed shape. The wind direction changes with the shape of the front.

## Summary

The sun provides the energy to make weather. The atmosphere traps heat near Earth's surface much as a greenhouse does. Air masses form over continents and oceans. When two air masses meet, they form a front. Fronts are the areas where most weather happens.

## Review

1. What is the greenhouse effect?
2. What is a weather front?
3. How does a cold front form?
4. **Critical Thinking** Why are weather forecasts sometimes incorrect?
5. **Test Prep** What kind of front forms when a warm air mass catches up to a cold air mass?

   **A** a warm front  **C** a rain front
   **B** a cold front   **D** a hot front

# LESSON 3

## How Is Weather Predicted?

In this lesson, you can . . .

 **INVESTIGATE** how to measure air pressure.

 **LEARN ABOUT** weather prediction.

 **LINK** to math, writing, drama, and technology.

◄ Rain falls from clouds when water droplets in the clouds become big and heavy.

D18

**INVESTIGATE**

# Air Pressure

**Activity Purpose** You've learned that air pressure is the force with which the atmosphere presses down on Earth. You've also learned that air pressure changes as weather changes. In this investigation you will make a barometer, an instrument to **measure** air pressure.

## Materials
- safety goggles
- scissors
- large, round balloon
- plastic jar
- large rubber band
- tape
- wooden craft stick
- small index card
- ruler

**CAUTION**

## Activity Procedure

1. **CAUTION** **Put on your safety goggles. Be careful when using scissors.** Use the scissors to carefully cut the neck off the balloon.

2. Have your partner hold the jar while you stretch the balloon over the open end. Make sure the balloon fits snugly over the jar. Secure the balloon with the rubber band.

3. Tape the craft stick to the top of the balloon as shown. Make sure that more than half of the craft stick stretches out from the edge of the jar. (Picture A)

Picture A

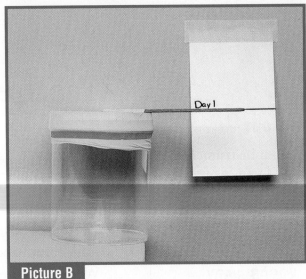

Picture B

**4** On the blank side of the index card, use a pencil and a ruler to make a thin line. Label the line *Day 1*. Tape the card to a wall. Make sure the line is at the same height as the wooden stick on your barometer. (Picture B)

**5** At the same time each day for a week, **measure** relative air pressure by marking the position of the wooden stick on the index card. Write the correct day next to each reading.

## Draw Conclusions

1. Describe how air pressure changed during the time that you were **measuring** it.

2. What might have caused your barometer to show little or no change during the time you were taking **measurements?**

3. **Scientists at Work** Meteorologists are scientists who use instruments to **measure** weather data. How did your barometer measure air pressure?

**Investigate Further** Use your air pressure **measurements** and information from daily weather reports to **predict** the weather in your area for the next few days.

**Process Skill Tip**

When you compare data to a standard, you are **measuring**. Careful measurements can help you make inferences or draw conclusions about your data.

# Weather Prediction

## Measuring Weather

Meteorologists (mee•tee•uhr•AHL•uh•jihsts) are scientists who study and measure weather conditions. Some of these conditions are air temperature, air pressure, and wind speed and direction. Meteorologists have developed tools for measuring each of these weather conditions.

When you want to know if it's hot or cold outside, you look at a thermometer or listen to a weather report. Thermometers measure the temperature of air. In the investigation, you measured another property of the atmosphere, air pressure. Air pressure is measured with an instrument called a **barometer** (buh•RAHM•uht•er). In one type of barometer, air presses down on the instrument, causing a needle to move. The needle points to a number that tells how much the air is pressing down. In other words, the instrument measures the weight of the air above it. You learned in Lesson 2 that the particles of air are closer together in a cold air mass than in a warm air mass. Most cold air masses are denser and so have higher air pressure than warm air masses.

This weather instrument is a rotating-drum barometer. It continuously records changes in air pressure. ▼

Earth's land surfaces heat faster than its bodies of water. So the air above land is usually warmer. That means it is also less dense. ▼

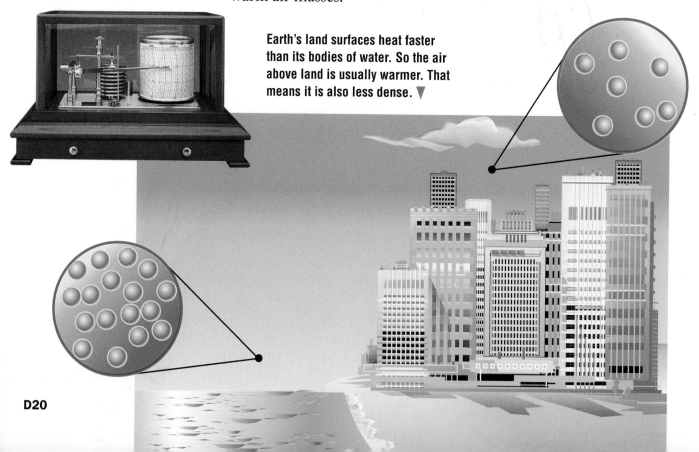

▲ Evaporating water adds more humidity to the air. Air masses that form over water have more moisture than those that form over land.

▲ A sling psychrometer (sy•KRAHM•uht•uhr) measures moisture in the air. The bulb of one thermometer is covered with a wet cloth. Then both thermometers are whirled in the air. The drier the air, the faster the water on the cloth evaporates (ee•VAP•uh•raytz), or dries up. This evaporation cools the cloth-covered thermometer. The temperatures of the wet and dry thermometers are compared to find the humidity.

Another characteristic of weather you can measure is **humidity** (hyoo•MID•uh•tee), or the amount of water vapor in the air. Humidity depends on several things. The area over which an air mass forms affects its humidity. For example, air masses that form over bodies of water have more moisture than air masses that form over land.

Temperature also affects how much moisture can be in the air. Warm air can have more water vapor than cool air. This is why water drops form on the outside of a glass of cold water during a warm day. Air near the glass cools. The air can no longer have as much water. The water vapor comes out of the air, forming drops.

In the investigation in Lesson 2, you made a windsock. With it you estimated the speed and direction of the wind. Meteorologists measure wind speed by using an instrument called an **anemometer** (an•uh•MAHM•uht•er). They find wind direction by using a weather vane or windsock.

✔ **What affects the humidity of air?**

◀ This is a type of anemometer that also includes a weather vane. Wind speed is measured by counting how many complete turns the cups make in one minute. Usually, a machine counts the turns.

| Daily Temperature Data for Weather City, Any State | | | |
|---|---|---|---|
| | Daily High/ Daily Low (°F) | Record High/ Record Low (°F) | Daily Average Temperature (°F) |
| Sunday | 89/72 | 101/44 | 81 |
| Monday | 90/74 | 100/50 | 82 |
| Tuesday | 85/65 | 99/42 | 75 |
| Wednesday | 88/69 | 103/40 | 79 |
| Thursday | 92/70 | 98/41 | 81 |
| Friday | 75/60 | 99/45 | 68 |
| Saturday | 79/62 | 102/44 | 71 |

▲ Charts like this are used to record daily weather conditions.

▲ Symbols on a weather map stand for fronts and weather conditions.

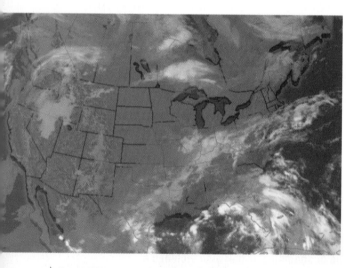

▲ A satellite photograph shows the positions of clouds and fronts. This satellite photograph matches the weather map above.

## Mapping and Charting Weather

By watching and measuring weather conditions, scientists can keep track of moving air masses. Scientists record their measurements on charts and maps. Then they analyze the data to predict weather.

A weather map can be big or small. Large maps show how the weather differs across a country. Small maps show weather changes across a state or a smaller area. Satellite pictures and maps often show clouds and weather for a large part of Earth.

A weather map uses symbols to show weather conditions. Long lines marked with half-circles or triangles stand for fronts. Words or symbols describe the weather in an area. For example, small dashes may stand for rain, and small stars may stand for snow. Symbols also may show the type of clouds that are in the area.

✔ **How do weather charts and maps help scientists predict the weather?**

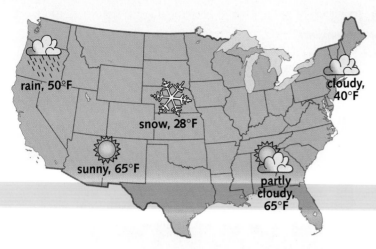

rain, 50°F

cloudy, 40°F

snow, 28°F

sunny, 65°F

partly cloudy, 65°F

▲ Simple weather maps like this are often printed in daily newspapers. What are the temperature and the weather in Nebraska?

## Summary

Meteorologists are scientists who study and measure weather conditions. These conditions include air pressure, air temperature, humidity, and wind speed and direction. By measuring and studying weather conditions, meteorologists are able to predict the weather.

## Review

1. What is an anemometer?

2. What factors affect humidity?

3. What are two tools that meteorologists use to study and predict the weather?

4. **Critical Thinking** An air mass forms over Alaska. Describe what you think the temperature and humidity of this air mass will be like.

5. **Test Prep** Which instrument is used to measure air pressure?

    A thermometer

    B weather vane

    C anemometer

    D barometer

# LINKS

### MATH LINK

**Temperature Differences** Make a chart like the one on page D22. In it, list the daily high and low temperatures of your area for one week. Find how much the temperature changed each day.

### WRITING LINK

**Expressive Writing—Friendly Letter** Suppose you have a pen pal who lives in an area of the country very different from your area. Write a letter to your pen pal. Describe how the weather in your area changes when a cold front moves through.

### DRAMA LINK

**Be a Weather Forecaster** Make a weather map showing imaginary weather conditions for your state. Present your forecast to the rest of the class. Make your presentation more interesting by using props.

### TECHNOLOGY LINK

Learn more about tools for measuring weather conditions by joining a *Tornado Chase* on **Harcourt Science Explorations CD-ROM.**

# Red Sprites, Blue Jets, and E.L.V.E.S.

**S**uppose you saw something completely new. How would you describe it? That was the challenge facing some airplane pilots and scientists. They tried to name unusual flashes they saw in the sky by calling them "upward lightning," "flames," and even "giant glowing doughnuts"!

## Sprites, Jets, and ELVES

In 1989 scientists began trying to show that these light flashes are real. A videotape showed the unusual flashes during a thunderstorm. Nearly 20 were photographed during the early 1990s using video cameras that could work with very little light.

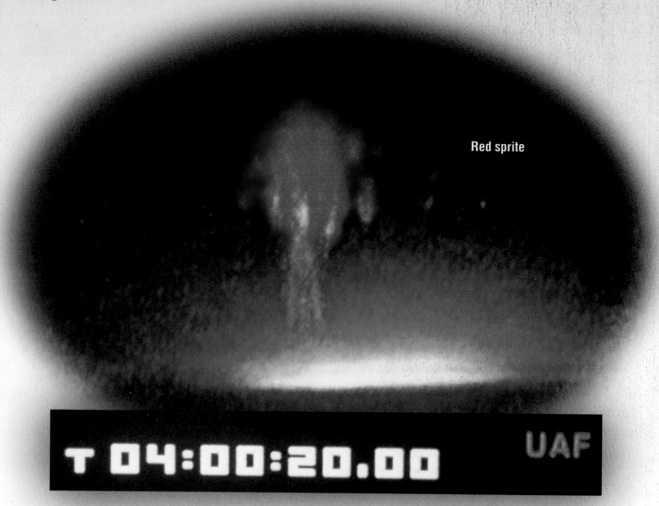

Red sprite

T 04:00:20.00   UAF

At least three types of flashes were identified. The scientists finally decided to name them sprites, jets, and ELVES.

**ELVES** stands for "*e*missions of *l*ight and *v*ery-low-frequency perturbations from *e*lectromagnetic-pulse *s*ources". You can see why the term is abbreviated. ELVES are very dim, quick red flashes. They move outward like ripples on a pond.

**Sprites** are red and seem to move in groups. They appear above a thunderstorm system, 65–75 kilometers (about 40–47 mi) above the ground. They have a "head" and strands coming down from the head. Sprites can occur over both sea and land.

**Blue jets** were once described as rocket lightning. Not until 1994 did weather scientists show, by using color video, that these glowing streaks are blue. Blue jets occur lower in the atmosphere than red sprites do, at 40–50 kilometers (about 25–30 mi) above the ground. Blue jets travel around 100 kilometers (about 62 mi) per second.

People in airplanes and on mountains could see blue jets and red sprites because the people were above storms. No one had ever seen ELVES because they last only a thousandth of a second. This is far too little time for the human eye to see. In 1990, videos taken from the space shuttle did show ELVES. Five more years passed before a second video of ELVES was made.

## Finding Sprites, Jets, and ELUES

There are several reasons why it took so long to discover these unusual lights:

- They occur only above thunderstorms, so clouds usually block the view from the ground.

- They are dim and can be seen only after the eyes have adjusted to the dark. That adjustment is spoiled by bright lightning.
- Sprites last only about 3 ten-thousandths of a second (0.0003 second).
- Only about 1 in 100 lightning strikes produces these lights.

It took careful observation to find out that sprites, jets, and ELVES are real.

## Think About It

**1.** What do you think scientists thought of the early reports of sprites and jets, before the videotapes?

**2.** Why do scientists use low-light video cameras to take pictures of sprites, jets, and ELVES?

**WEB LINK:**
**For Science and Technology updates, visit the Harcourt Internet site.**
**www.harcourtschool.com**

## Careers    Meteorologist

**What They Do**
Meteorologists study the atmosphere and the changes that produce different kinds of weather. Meteorologists may work for business or government. They may also research new uses for computer programs in the study of weather.

**Education and Training**   Someone who wants to be a meteorologist must study science in college. Many meteorologists get further training in weather research and technology after college.

# Denise Stephenson-Hawk

## ATMOSPHERIC SCIENTIST

Dr. Denise Stephenson-Hawk always loved school and was especially good in math. She skipped her senior year in high school and entered Spelman College. While there, she received a scholarship to a summer program at the National Aeronautics and Space Administration (NASA). She worked on a project to test panels for the space shuttle to make sure the panels would withstand the high temperatures they experience upon reentry into Earth's atmosphere. Stephenson-Hawk was so excited by what she learned about the atmosphere that she decided to apply her math skills to the study of atmospheric science.

Stephenson-Hawk's first job was at AT&T Bell Laboratories. There she made computer models to learn how sound travels in the ocean. After teaching mathematical modeling at Spelman College, Stephenson-Hawk moved to Clark Atlanta University in Georgia. There she works as a senior research scientist and associate professor of physics.

Stephenson-Hawk is also a member of the Climate Analysis Center (CAC) at the National Oceanic and Atmospheric Administration (NOAA). The CAC uses computer models to

analyze and predict climate changes that happen in a short time. Stephenson-Hawk's special project has been building computer models of the effects of El Niño, a series of events set off by warmer-than-normal surface-water temperatures in the Pacific Ocean. Stephenson-Hawk and the other scientists working on this project are trying to more accurately predict the impact of El Niño so that people can better prepare for unusual weather.

## THINK ABOUT IT

1. Why else might scientists be interested in studying El Niño?

2. Why do you think atmospheric scientists use computer models?

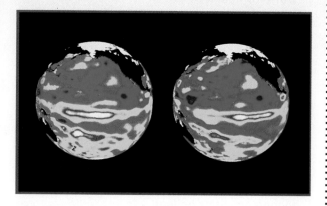

**El Niño water temperature maps**

# Relative Wind Speed

## What is a way to measure relative wind speed?

### Materials

- pattern (TR p.107)
- cardboard
- scissors
- permanent marker
- plastic straw
- long pencil
- masking tape
- paper fastener
- small paper cup

### Procedure

1. **CAUTION** **Be careful when using scissors.** Trace the pattern pieces onto the cardboard and cut them out. Add the scale markings.

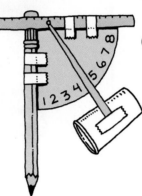

2. Cut a hole for the pencil in the middle of the straw. Push the pencil eraser into the hole. Tape the straw and pencil to the corner of the cardboard wedge.

3. Use the fastener to attach the cardboard strip. Tape the cup to the strip.

4. Push the pencil point into the ground in an open, windy area. Observe and record relative wind speed twice a day for one week. Use the table on page D11 to help you match gauge readings to actual wind speeds.

### Draw Conclusions

How did the gauge help you measure wind speed?

# Weather Fronts

## How can water model a weather front?

### Materials

- tall, clear jar
- hot and cold tap water
- pitcher
- food coloring
- thermometer

### Procedure

1. Fill the jar halfway with cold water.

2. Fill the pitcher with hot water. Add 10 drops of food coloring.

3. Tilt the jar of cold water. Then slowly trickle the hot water down the inside of the jar. Slowly put the jar upright. Observe what happens in the jar.

4. Use the thermometer to measure the temperature of the hot water in the jar. Carefully move the thermometer down to measure the cold water in the jar. Can you find the front by using the thermometer?

### Draw Conclusions

How did the hot water and cold water interact? How were they like air masses?

# Chapter 1 Review and Test Preparation

## Vocabulary Review

Use the terms below to complete the sentences. The page numbers in ( ) tell you where to look in the chapter if you need help.

**atmosphere** (D6)

**air pressure** (D7)

**troposphere** (D8)

**stratosphere** (D8)

**greenhouse effect** (D12)

**air mass** (D13)

**front** (D14)

**barometer** (D20)

**humidity** (D21)

**anemometer** (D21)

1. The _____ is the thin layer of air that surrounds Earth.

2. The amount of water vapor in the air is called _____.

3. The warming caused when air traps some of the sun's energy is the _____.

4. _____ is the force with which the atmosphere presses down on Earth.

5. In the atmosphere, the _____ is the layer in which most weather occurs.

6. An _____ is a large body of air, and it forms and moves over land or water.

7. An instrument that measures air pressure is a _____.

8. A _____ forms when two air masses meet.

9. In the atmosphere, the layer that contains a lot of ozone is the _____.

10. An instrument that measures wind speed is an _____.

## Connect Concepts

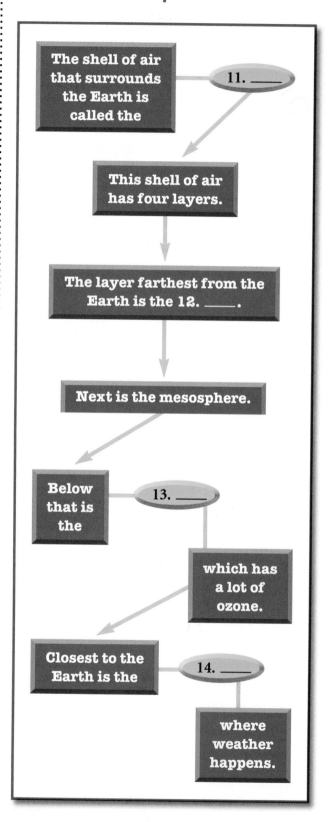

The shell of air that surrounds the Earth is called the

11. _____

This shell of air has four layers.

The layer farthest from the Earth is the 12. _____.

Next is the mesosphere.

Below that is the

13. _____

which has a lot of ozone.

Closest to the Earth is the

14. _____

where weather happens.

## Check Understanding

Write the letter of the best choice.

15. As you get higher in the atmosphere, the space between air particles —
    A decreases
    B doesn't change
    C increases
    D masses

16. Energy from the _____ is trapped by gases in the air, causing the greenhouse effect.
    F Earth        H barometer
    G sun          J stratosphere

17. An air mass that forms over tropical waters would be —
    A warm and moist
    B cold and moist
    C cold and dry
    D warm and dry

18. A _____ front forms when two air masses meet and don't move.
    F cold         H stationary
    G warm         J pressure

19. _____ air can have more water vapor than _____ air.
    A Warm, cold
    B Dense, less dense
    C Cold, warm
    D Thermosphere, mesosphere

## Critical Thinking

20. Why do mountain climbers use oxygen tanks?

21. You hear on a weather report that a cold front is coming. What weather changes can you expect?

22. Suppose you watch the weather report each day for five days. Each day the average temperature is the same and the air pressure doesn't change. What could be happening?

## Process Skills Review

23. Remember the first investigation in this chapter. What did you **observe** that allowed you to **infer** that air is matter?

24. How is a **measurement** of wind speed different from a word description of wind speed?

25. Suppose you will **measure** weather conditions over the next five days. What equipment will help you measure? Make a table to record your data. Include units of measure.

## Performance Assessment

**Weather Maps**

With a partner, study the three maps your teacher gives you. Tell how the weather has changed in the map area over the past three days. Then predict what the weather will be for the next two days. Explain the reasons for your prediction.

# Water in the Oceans

**N**early three-fourths of Earth is covered with a great ocean of salt water. It is a moving body of water and full of life. Its currents bring warm temperatures to otherwise cold areas. Its depths hide great mountain ranges. And its nutrient-rich waters are home to all sorts of living things.

## Vocabulary Preview

water cycle
evaporation
condensation
precipitation
wave
storm surge
tide
deep ocean current
surface current

### FAST FACT

The oceans of the Earth are vast and deep. If Earth were a smooth ball with no mountains or valleys at all, it would be completely covered with water to a depth of more than 2 kilometers (about $1\frac{1}{4}$ mi).

Mount Everest

The deepest spot in the ocean is in the Mariana Trench in the Pacific—11,000 meters (about 36,000 ft) below sea level. If Mount Everest, Earth's highest mountain, were dropped into that spot, it would be covered with about $1\frac{1}{2}$ kilometers (about 1 mi) of water!

FAST FACT

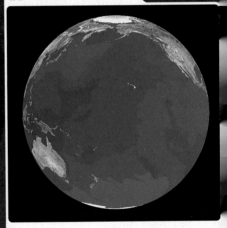

The Pacific Ocean holds about half of Earth's ocean water, and it covers nearly a third of Earth's surface. Here's how three oceans compare:

## Ocean Sizes

| Ocean | Size (square kilometers) | Size (square miles) |
|---|---|---|
| Pacific | 181,000,000 | 70,000,000 |
| Atlantic | 94,000,000 | 36,000,000 |
| Indian | 74,000,000 | 29,000,000 |

# What Role Do Oceans Play in the Water Cycle?

In this lesson, you can . . .

**INVESTIGATE** how to get fresh water from salt water.

**LEARN ABOUT** Earth's ocean water.

**LINK** to math, writing, social studies, and technology.

◀ Buoys float but are held in place by anchors. They mark paths where the water is deep enough for ships.

## INVESTIGATE

# Getting Fresh Water from Salt Water

**Activity Purpose**  If you've ever been splashed in the face by an ocean wave, you know that sea water is salty. The salt in ocean water stings your eyes, leaves a crusty white coating on your skin when it dries, and tastes like the salt you put on food. In this investigation you'll evaporate artificial ocean water to find out what is left behind. From your **observations** you will **infer** how you can get fresh water from salt water.

## Materials

- container of very warm water
- salt
- spoon
- cotton swabs
- large clear bowl
- small glass jar
- plastic wrap
- large rubber band
- piece of modeling clay
- masking tape

**CAUTION**

## Activity Procedure

**1** Stir two spoonfuls of salt into the container of very warm water. Put one end of a clean cotton swab into this mixture. Taste the mixture by touching the swab to your tongue. **Record** your observations. **CAUTION** **Don't share swabs. Don't put a swab that has touched your mouth back into any substance. Never taste anything in an investigation or experiment unless you are told to do so.**

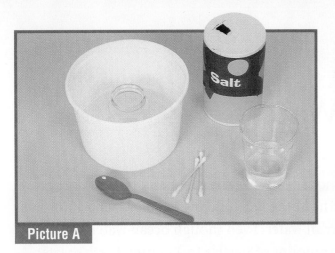

Picture A

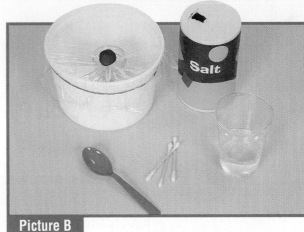

Picture B

2️⃣ Pour the salt water into the large bowl. Put the jar in the center of the bowl of salt water. (Picture A)

3️⃣ Put the plastic wrap over the top of the bowl. The wrap should not touch the top of the jar inside the bowl. Put a large rubber band around the bowl to hold the wrap in place.

4️⃣ Form the clay into a small ball. Put the ball on top of the plastic wrap right over the jar. Make sure the plastic wrap doesn't touch the jar. (Picture B)

5️⃣ On the outside of the bowl, use tape to mark the level of the salt water. Place the bowl in a sunny spot for one day.

6️⃣ After one day, remove the plastic wrap and the clay ball. Use clean swabs to taste the water in the jar and in the bowl. **Record** your **observations.**

## Draw Conclusions

1. What did you **observe** by using your sense of taste?

2. What do you **infer** happened to the salt water as it sat in the sun?

3. **Scientists at Work**  The movement of water from the Earth's surface, through the atmosphere, and back to Earth's surface is called the water cycle. From what you **observed**, what can you **infer** about the ocean's role in the water cycle?

**Investigate Further**  Put the plastic wrap and the clay back on the large bowl. Leave the bowl in the sun for several days, until all the water in the large bowl is gone. **Observe** the bowl and the jar. What can you **conclude** about ocean water?

**Process Skill Tip**

Observing and inferring are different things. You **observe** with your senses. You **infer**, or form an opinion, based on what you have observed and what you know about a situation.

# Ocean Water

## The Water Cycle

**FIND OUT**

- about processes that make up the water cycle
- why ocean water is salty

**VOCABULARY**

water cycle
evaporation
condensation
precipitation

Oceans cover more of Earth's surface than dry land does. About three-fourths of the Earth is covered by water. Almost all of that water is ocean water. Even though ocean water is salty, it provides a large amount of Earth's fresh water. Earth's water is always being recycled. As the model in the investigation showed, heat from the the sun causes fresh water to evaporate (ee•VAP•uh•rayt) from the oceans, leaving the salt behind. This evaporated water condenses to form clouds. Fresh water falls from the clouds to Earth's surface as rain. This constant recycling of water is called the **water cycle**. During the cycle, water changes from a liquid to a gas and back to a liquid. The diagram on these pages shows how the water cycle works. It includes the parts played by the sun, the water, the air, and the land.

✔ **What is the water cycle?**

The sun warms the ocean, causing the water particles to move faster and faster. After a while, they have enough energy to leave the water and enter the air as water vapor. This is **evaporation**, the process by which a liquid changes to a gas. ▼

A cloud forms when water vapor condenses high in the atmosphere. **Condensation** (kahn•duhn•SAY•shuhn) happens when the water vapor rises, cools, and changes from a gas to liquid water. These drops of water in a cloud are so small that they stay up in the air.

Water vapor from an ocean can be carried a long way through the atmosphere. Water that evaporates from the Gulf of Mexico may fall back to Earth's surface far away in North Carolina.

In the cloud, some water drops bump into others and stick together. The drops get bigger. When they get too large to stay up in the air, they fall to Earth as rain, snow, sleet, or hail. Some of this **precipitation** (pree•sip•uh•TAY•shuhn) collects in lakes, rivers, and other bodies of water. Some precipitation soaks into the ground to become groundwater. Some falls directly back into the ocean.

# What Is in Ocean Water

Ocean water is a mixture of water and many dissolved solids. Most of these solids are salts. Sodium chloride is the most common salt in ocean water. You probably know this substance by another name—table salt.

Where do you think the salts and other solids in the ocean come from? Most of the salts and other substances in the ocean come from the land. As rivers, streams, and runoff flow over the land, they slowly break down the rocks that make it up. Over time, flowing water carries substances from the rocks to the ocean.

**Ocean Water Content**

4% salts and other solids

96% water

▲ Ocean water is made up of almost the same substances everywhere on Earth. Ocean water is about 96 percent water and 4 percent salts and other dissolved solids.

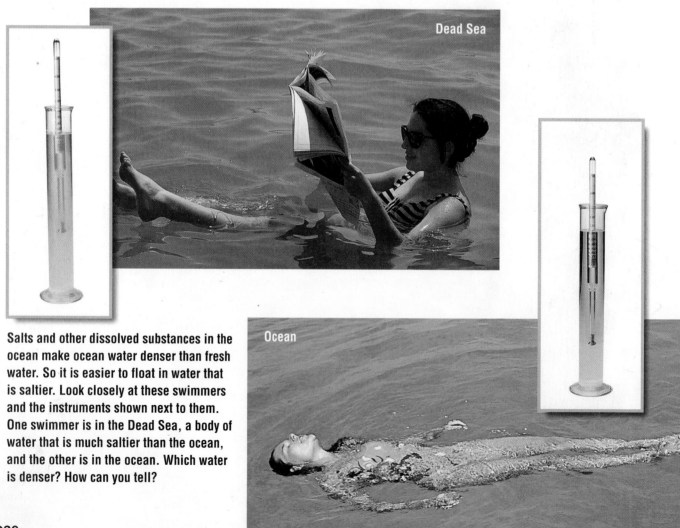

Dead Sea

Ocean

Salts and other dissolved substances in the ocean make ocean water denser than fresh water. So it is easier to float in water that is saltier. Look closely at these swimmers and the instruments shown next to them. One swimmer is in the Dead Sea, a body of water that is much saltier than the ocean, and the other is in the ocean. Which water is denser? How can you tell?

Near places where rivers empty into the ocean, the ocean water is less salty than it is farther from the shore. This is because the fresh water mixes with the salt water. Ocean water is a little saltier near the equator, where it is hot and water evaporates faster. And ocean water is a little less salty near the North and South Poles, where it is colder and water evaporates more slowly.

✔ **What is in ocean water?**

## Summary

The waters of the ocean provide fresh water for Earth through the water cycle. As water moves through this cycle, it changes from a liquid to a gas and back to a liquid through the processes of evaporation and condensation. The water returns to Earth as precipitation. Sodium chloride is the most common salt in the ocean. The salts and other substances dissolved in ocean water make it denser than fresh water.

## Review

1. What is the water cycle?
2. Explain how water changes from a liquid to a gas and back to a liquid in the water cycle.
3. What factors affect the saltiness or density of ocean water?
4. **Critical Thinking** How could you make salt water denser?
5. **Test Prep** Which of these processes occurs when a gas changes to a liquid?
   - **A** evaporation
   - **B** condensation
   - **C** precipitation
   - **D** salinity

# LINKS

### MATH LINK

**Compare Fresh Water and Salt Water** Use library reference materials to find out more about the amounts of fresh water and salt water on Earth. Draw a large square on a sheet of paper, and divide it into fourths. Color the squares to show the amounts of land and ocean. Stack pennies or checkers on the squares to stand for the amounts of fresh water and salt water.

### WRITING LINK

**Narrative Writing—Story** Suppose you are sailing alone around the world. For your teacher, write down some of your thoughts that describe the ocean and what it is like to have nothing but water all around you.

### SOCIAL STUDIES LINK

**El Niño** Find out what El Niño is. Locate on a world map the places where this condition occurs. Write a report that explains what causes this situation and how it affected weather and crops around the world in 1998.

### TECHNOLOGY LINK

Learn more about Earth's water systems by visiting the National Air and Space Museum Internet site.
**www.si.edu/harcourt/science**

Smithsonian Institution®

LESSON **2**

# What Are the Motions of Oceans?

In this lesson, you can . . .

 **INVESTIGATE** water currents.

 **LEARN ABOUT** the ways ocean water moves.

 **LINK** to math, writing, social studies, and technology.

**INVESTIGATE**

# Water Currents

**Activity Purpose** If you've ever gone swimming in the ocean, you've probably felt waves crash against your body. You may also have felt water moving against you below the surface. This movement below the water's surface is a *current*. In this investigation you'll **make a model** and **infer** one way currents form.

### Materials
- clear, medium-sized bowl
- warm tap water
- colored ice cube
- clock

## Activity Procedure

**1** Put the bowl on a flat surface. Carefully fill the bowl three-quarters full of warm tap water.

**2** Let the water stand undisturbed for 10 minutes.

◄ Ocean water moves in many ways. Both the rising water and the waves are washing away this sand castle.

**3** Without stirring the warm water or making a splash, gently place the colored ice cube in the middle of the bowl. (Picture A)

**4** **Observe** for 10 minutes what happens as the ice cube melts. Every 2 minutes, make a simple drawing of the bowl to **record** your observations.

**Picture A**

## Draw Conclusions

1. Describe what you **observed** as the ice cube melted in the bowl of warm water.

2. In your **model**, what does the bowl of water stand for? What does the ice cube stand for?

3. Since the liquid in the bowl and the ice cube were both water, what can you **infer** about the cause of what happened in the bowl?

4. **Scientists at Work** In Chapter 1, you learned that cold air is denser than warm air. The same is true for water. Using this information and what you **observed** in the investigation, explain one way ocean currents form.

**Investigate Further** Mix up two batches of salt water. Use twice as much salt in one batch as in the other. Use the water to **model** another kind of ocean current. Fill a clear bowl three-fourths full with the less salty water. Add a few drops of food coloring to the saltier water. Along the side of the bowl, slowly pour the colored, saltier water into the clear, less salty water. Describe your **observations. Make a hypothesis** to explain what you observed. What **prediction** can you make based on the hypothesis? How could you test the prediction?

---

**Process Skill Tip**

People make models to help them **observe** things in nature that are too small, too big, or too hard to see or understand. By observing a model, you can infer how things work.

# Ocean Movements

**FIND OUT**

• about ocean waves and currents

• what causes tides

**VOCABULARY**

wave
storm surge
tide
surface current
deep ocean current

## Waves

If you've ever been in the path of a wave in the ocean, in a lake, or in a wave pool, you know that even big waves don't move you either forward or back. You bob up and down, but you're still in about the same place after the wave passes. This is because a **wave** is the up-and-down movement of the water particles that make it up.

Water waves are caused by the wind. As wind blows over the water's surface, it pulls on the water particles. This causes small bumps, or ripples, of water to form. As the wind continues to blow, the ripples keep growing. Over time they become waves.

The height of a wave depends on three things: the strength of the wind, the amount of time the wind blows, and the size of the area over which the wind blows. Strong, gusty winds blowing

Waves drop and take away bits of rock and sand grains from a beach as they break on a rocky shore. ▶

Waves like these are caused by the wind. Waves break, or give up their energy, as they move onto the shore. ▼

▲ Waves erode a shore as the water carries sediment back toward the sea.
What might happen to these houses if erosion continues?

over an area of many square kilometers can cause a very large wave called a storm surge to form. Storm surges often occur during hurricanes and can cause a lot of damage along a shore.

Waves change the shore in different ways. When waves break on a beach, water carries sand and other sediments as it flows back into the ocean. This carrying away of sediments is called *erosion*. Erosion along a shore causes beaches to become smaller. As waves give up their energy, they also deposit, or drop, sediments. This process is called

*deposition* (dep•uh•ZISH•uhn). When waves deposit sediments near shore, a beach gets bigger.

The photograph at the bottom of this page shows a harbor during a hurricane. You probably know that a hurricane is a severe storm that has strong winds and a lot of rain. Storm surges during hurricanes cause erosion and deposition along a shore. Whole beaches can be washed away.

✓ **How do water waves change a shoreline?**

This is a harbor during a hurricane. Storm surges during hurricanes can be as high as 10 meters (more than 30 ft). ▼

D41

## Tides

If you watched a beach for 12 hours, you would probably notice that the waves don't always reach the same place. This is because of another type of ocean water motion called tides. **Tides** are the daily change in the local water level of the ocean.

At *high tide* much of the beach is covered by water. At *low tide* waves break farther away from shore. Less of the beach is under water. Every day most shorelines have two high tides and two low tides. High tide and low tide are usually a little more than six hours apart.

Tides are caused by gravity. *Gravity* is a force that causes all objects to be pulled toward all other objects. The force of gravity between two objects depends on two things: the sizes of the objects and the distance between them. Big objects have a greater pull than small objects. Objects that are closer together have a greater pull on one another than objects farther apart do.

Even though the moon is much smaller than the sun, the pull of the moon's gravity on Earth is the main cause of ocean tides. This is because the moon is much closer to Earth than the sun is.

▲ This photograph shows low tide in a harbor on the Bay of Fundy in Nova Scotia, Canada.

▲ This is the same harbor on the Bay of Fundy during high tide. Compare the positions of the ships with their positions in the photograph above.

◄ A tide pool is a small body of water on the shore. Parts of it may be out of the water during low tide. Sea animals such as starfish, crabs, and sea anemones (uh•NEM•uh•neez) live in these pools.

The moon pulls on everything on Earth. As the moon pulls on ocean water, the water forms a bulge that always faces the moon. Another water bulge forms on the side of Earth farthest from the moon, where the moon's pull is weakest. As Earth rotates, the bulges stay in the same places. High tide occurs as a point on Earth moves through a bulge. The water level rises on the shore. Low tide occurs as a point on Earth moves between the bulges. The water level on the shore gets lower.

At certain times each month, tides are very high or very low. Read The Inside Story to find out why.

✓ **What are tides?**

# THE INSIDE STORY

## The Moon and Tides

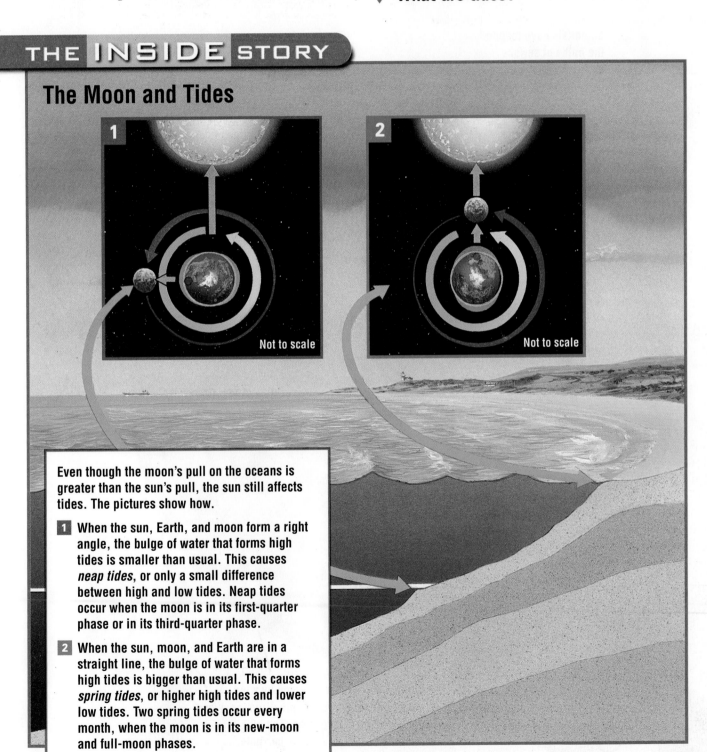

Not to scale

Not to scale

Even though the moon's pull on the oceans is greater than the sun's pull, the sun still affects tides. The pictures show how.

**1** When the sun, Earth, and moon form a right angle, the bulge of water that forms high tides is smaller than usual. This causes *neap tides*, or only a small difference between high and low tides. Neap tides occur when the moon is in its first-quarter phase or in its third-quarter phase.

**2** When the sun, moon, and Earth are in a straight line, the bulge of water that forms high tides is bigger than usual. This causes *spring tides*, or higher high tides and lower low tides. Two spring tides occur every month, when the moon is in its new-moon and full-moon phases.

# Currents

*Currents* are rivers of water that flow in the ocean. A **surface current** forms when steady winds blow over the surface of the ocean. In the Northern Hemisphere, surface currents flow in a clockwise direction. In the Southern Hemisphere, surface currents flow in a counterclockwise direction.

**Deep ocean currents** form because of density differences in ocean water. You made a model of deep ocean currents in the investigation. The density of ocean water depends on two things—the amount of salt in the water and the temperature of the water. The more salt there is in water, the denser it will be. Cold ocean water also is

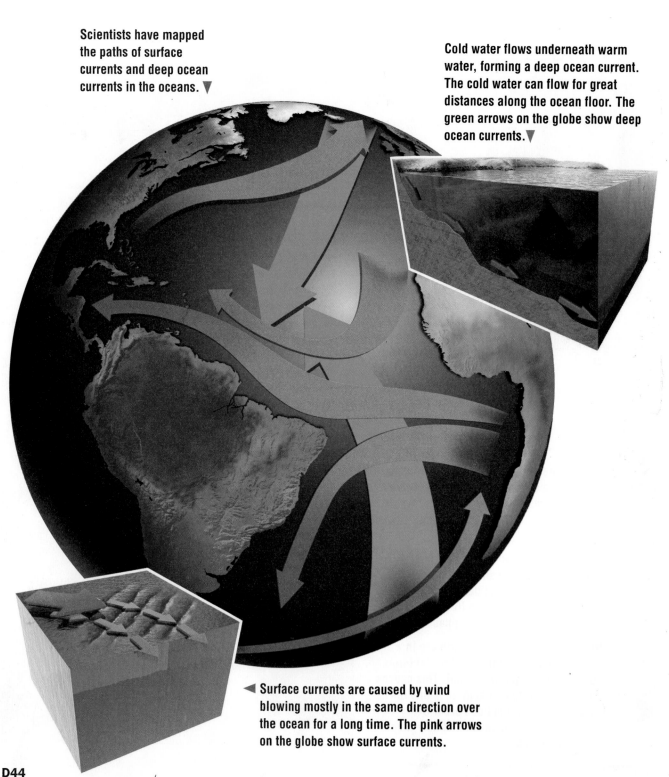

Scientists have mapped the paths of surface currents and deep ocean currents in the oceans. ▼

Cold water flows underneath warm water, forming a deep ocean current. The cold water can flow for great distances along the ocean floor. The green arrows on the globe show deep ocean currents. ▼

◄ Surface currents are caused by wind blowing mostly in the same direction over the ocean for a long time. The pink arrows on the globe show surface currents.

denser than warm ocean water. Deep ocean currents form when dense, cold water meets less dense water. The denser water flows under the less dense water, forcing the less dense water to rise.

✓ **What causes deep ocean currents?**

## Summary

Ocean waves form as wind blows over the water's surface. Strong, gusty winds over a large area can cause a large wave called a storm surge. Tides are the rise and fall of ocean water caused by the pull of gravity between Earth, the moon, and the sun. Surface currents are caused by the blowing of steady winds. Deep ocean currents are caused by differences in saltiness or water temperature.

## Review

1. How does an object floating in water move as a wave passes it?
2. What causes tides?
3. Why does the moon have a greater effect on tides than the sun has?
4. **Critical Thinking** Compare and contrast surface currents in the Northern and Southern Hemispheres.
5. **Test Prep** Which is caused by density differences?
   A surface currents
   B ocean waves
   C deep ocean currents
   D ocean tides

# LINKS

## MATH LINK

**Rising Sea Level** Scientists estimate that sea level rose 10 to 15 centimeters between the years 1900 and 1998. They estimate that it will rise another 30 centimeters before 2025. What will the average yearly rate of sea level rise be for the years 1998 to 2025?

## WRITING LINK

**Informative Writing—Description** Suppose you are a creature that lives in a tide pool. For a younger student, write a short description that explains how your life changes when the tide comes in and goes out.

## SOCIAL STUDIES LINK

**Trade Routes** Find out about the triangle trade route that existed in colonial times in the United States. Mark this route on a copy of a world map, and explain how ocean currents and wind patterns made this route possible.

## TECHNOLOGY LINK

Learn more about new ways to explore deep in the ocean by viewing *Robot Submarine* on the **Harcourt Science Newsroom Video.**

# Deep Flight II

**D**eep Flight II is a submersible, or sub, that is being developed to explore the ocean depths. It will carry a pilot and will go to the deepest part of the ocean, the Mariana Trench. This trench in the Pacific Ocean is 11,275 meters (almost 7 mi) deep. From that depth, it would take more than 25 Empire State Buildings stacked on top of each other to reach the surface.

Graham Hawkes, an engineer, and his business partner, Sylvia Earle, have been working to make deep-ocean subs, such as *Deep Flight II*. Scientists want better subs to learn more about the ocean. The subs will help them investigate ocean-floor geology as well as deep-ocean plants and animals. Subs, however, can't meet all research

## Exploring the Deep Ocean

Only one mission with people has ever gone to the Mariana Trench. In 1960 two men in a bathyscaph (BATH•uh•skaf) called the *Trieste* (tree•EST) went down for 20 minutes. The ship had none of the video cameras or computers we have today. And the *Trieste* could only go straight down and come straight back up.

**Graham Hawkes in a museum model of *Deep Flight I***

needs. Robot, or remote-controlled, subs are often better for dangerous or long trips.

## Built for Speed and Comfort

The time needed to go so deep is a problem, so Hawkes is designing a craft that "flies" through water like an airplane. It should reach the ocean floor in 90 minutes.

Another problem is the high pressure. The sub must support the weight of a column of water 11,275 meters high, so the hull of *Deep Flight II* is being built of strong, new ceramic materials. These materials are lighter than steel and won't break under high pressure. The ship will also protect its crew from the near-freezing cold of the ocean water.

Hawkes is using shapes from nature for the design of *Deep Flight II*. Its smooth body and wings look like parts of dolphins, whales, and birds, as well as aircraft. It can skim forward below the ocean's surface or dive down into the depths. It can even do spins and rolls like a stunt airplane.

*Deep Flight II* has an unusual feature. The pilot is strapped face down in a form-fitting body pan. He or she can see out through a cone-shaped window at the front. Hawkes explains that this is a natural swimming position, so it feels right to the pilot.

If the first voyage of *Deep Flight II* is a success, Hawkes and Earle may soon be designing and building more deep-sea subs.

## Think About It

1. How will the pilot of *Deep Flight II* be like an astronaut landing on the moon?
2. Why do you think most of the ocean remains unexplored?

**WEB LINK:**
**For Science and Technology updates, visit the Harcourt Internet site.**
**www.harcourtschool.com**

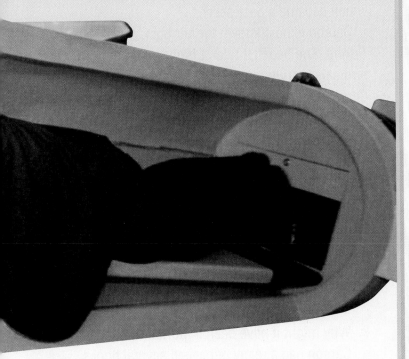

| **Careers** | **Scuba Support Crew** |

### What They Do

People working on a scuba support crew prepare equipment for diving. They make sure that everything needed for a dive *works.* The scuba support crew stays on land or in a boat while the diver goes under water. The crew communicates with the diver and takes care of any emergencies.

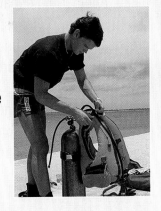

**Education and Training**   Scuba support crew members must have emergency medical training and a Divemaster certificate from the Professional Association of Diving Instructors.

# Rachel Carson
## MARINE BIOLOGIST

*"Science is part of the reality of living: it is the what, the how, and the why of everything in our experience."*

Although Rachel Carson did not actually visit the ocean until she was 22 years old and a college graduate, she had been fascinated by it all her life.

Everything in nature thrilled Carson—flowers and birds, trees and rivers, animals and insects. She surprised many people at her college by changing from an English major to a biology major. At the time, science was seen as a career for men. No one expected her to do well.

Carson used both biology and writing in her work. She taught college classes after she graduated. Another of her early jobs was with the U.S. Bureau of Fisheries, writing scripts for radio broadcasts about life under the sea. She eventually moved up in the Bureau to become editor-in-chief of publications.

Carson was encouraged to write articles, which eventually were printed in one book, *Under the Sea-Wind.* Carson wrote other books about the sea, including *The Sea Around Us,* which became a best-seller.

Carson is best known for her book *Silent Spring.* She began it after a friend who had a bird sanctuary wrote a letter. Her friend wrote Carson that pesticides had killed all the birds at the sanctuary. It took Carson several years to collect information and write the book. She closely studied the dangers of DDT, a chemical used to kill insects. *Silent Spring* tells what a spring would be like without new life.

## THINK ABOUT IT

1. Why was it important that Carson spend so much time collecting information for *Silent Spring*?

2. Why might it have been important that Carson had already published several books before writing *Silent Spring*?

Spiny sea urchins

## Measuring Density

### What are the relative densities of different solutions?

**Materials**

- unsharpened pencil with eraser
- tall, narrow glass or jar
- permanent marker
- safety goggles
- thumbtack
- water
- salt
- spoon

**Procedure**

1. **CAUTION** **Put on your goggles.** Carefully push the thumbtack completely into the eraser.

2. Fill the glass three-fourths full of tap water. Place the pencil, eraser end first, into the water. When it floats up on its own, grab the pencil at the point where it meets the surface of the water. Use the permanent marker to mark this point.

3. Add four spoonfuls of salt to the water. Stir until the salt dissolves. Place the pencil in the water, and mark it as you did in Step 2.

**Draw Conclusions**

Explain your observations in Steps 2 and 3.

## Making Waves

### How does a wave move?

**Materials**

- safety goggles
- heavy washer
- 2-m length of rope

**Procedure**

1. **CAUTION** **Put on your goggles.** String the washer onto the rope. It should fit loosely.

2. Work with a partner. Each of you should hold one end of the rope. It should hang loosely with the washer in the center of the rope. The rope should not be stretched between you and your partner.

3. Shake one end of the rope by moving your arm while your partner holds the other end still. Observe the movement of the washer on the rope. Compare this movement to the movement of objects floating in ocean water.

**Draw Conclusions**

How did the washer move? How is the movement of the rope like the movement of a wave?

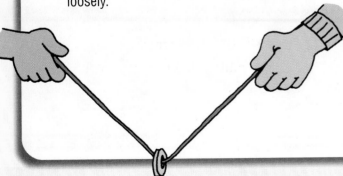

# Chapter ② Review and Test Preparation

## Vocabulary Review

Use the terms below to complete the sentences. The page numbers in ( ) tell you where to look in the chapter if you need help.

evaporation (D34)   storm surge (D40)

water cycle (D34)   tides (D42)

condensation (D35)   surface current (D44)

precipitation (D35)   deep ocean

wave (D40)   current (D44)

1. A large wave caused by strong winds is called a ___.

2. ___ is a process that changes a liquid to a gas.

3. A gas changes to a liquid by a process called ___.

4. Blowing winds form a ___, a river of water in the ocean.

5. Water vapor, clouds, rain, and the ocean are parts of the ___.

6. Neap and spring ___ are examples of changes in ocean level caused by the pull of the sun, moon, and Earth.

7. ___ is any form of water that falls from clouds.

8. A ___ forms when cold, dense ocean water meets and flows beneath warmer ocean waters.

9. The up-and-down motion of water is called a ___.

## Connect Concepts

The diagram below shows how water moves from the land and oceans to the atmosphere and back again. Label each section of the diagram, and write a title for it.

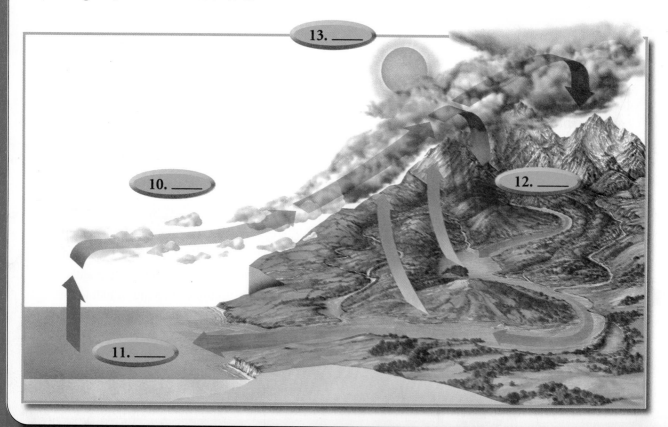

13. ___

10. ___

12. ___

11. ___

## Check Understanding

Write the letter of the best choice.

**14.** The _____ provides the energy for the water cycle.

    **A** moon     **C** atmosphere

    **B** sun       **D** ocean wave

**15.** Ocean water is a mixture of water and many —

    **F** gases     **H** liquids

    **G** salts     **J** fluids

**16.** Wind causes an up-and-down movement of ocean water, called a —

    **A** deep ocean current

    **B** tide

    **C** surface current

    **D** wave

**17.** Waves change a shoreline by erosion and —

    **F** condensation

    **G** precipitation

    **H** deposition

    **J** evaporation

**18.** The pull of _____ on Earth is the main reason for tides.

    **A** the moon

    **B** winds

    **C** the sun

    **D** currents

## Critical Thinking

**19.** Would it be easier to float in the Great Salt Lake or in Lake Michigan? Explain your answer.

**20.** What do you think would happen to tides if the moon's gravity had a stronger pull than it does now?

## Process Skills Review

**21.** Suppose there is a heavy rainstorm far out over the ocean. What can you **infer** about the density of the ocean water at the surface in that area right after the storm?

**22.** You **observe** clouds forming on a bright, sunny day. What can you **infer** is happening in the atmosphere? What may happen later in the day?

## Performance Assessment

### Currents

Completely fill a plastic cup with hot water. Add three or four drops of food coloring to the water. Cover the cup with plastic wrap. Hold the wrap in place with a rubber band.

Put the cup inside a bowl. Fill the bowl with cold water until it is almost full. There should be 2–3 cm of water over the cup. Use a pencil to poke a hole in the plastic wrap. Observe what happens. Explain what is happening inside the bowl. How could you make the same thing happen using salt water instead of hot water?

# Planets and Other Objects in Space

From Earth you can study objects in space by just stepping outside on a clear night. Most of the objects you will see are stars, which are very, very distant suns. A few of the objects you will see are planets. Some are a little like Earth, and some are amazingly different.

## Vocabulary Preview

solar system
star
planet
asteroid
comet
orbit
axis
inner planet
outer planet
gas giant
telescope
space probe
constellation

**FAST FACT**

The sun is about 150,000,000 kilometers (93 million mi) from the Earth. It would take you about 193 years to travel this distance in a car at highway speed!

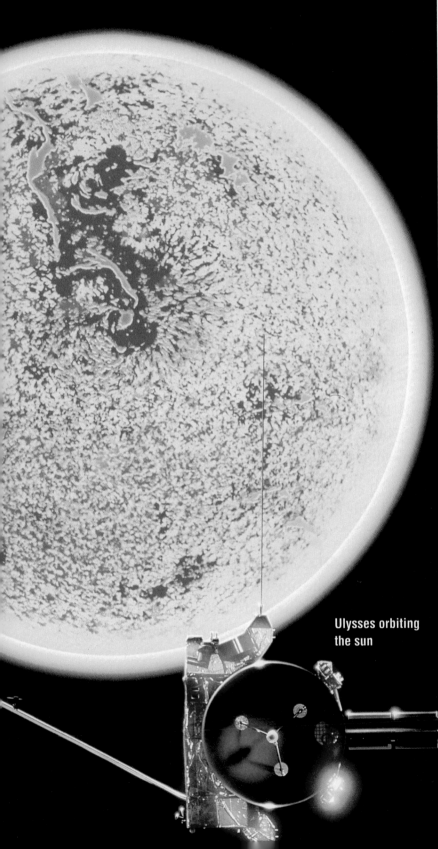

Ulysses orbiting the sun

About 500,000 craters can be seen on the moon through telescopes on Earth. It would take you more than 400 hours to count them all. And this doesn't include the craters on the far side of the moon!

## FAST FACT

Like most of the planets, Earth has seasons because it is tilted on its axis. But no planet is tilted like Uranus. Uranus is tilted so far that it is tipped over on its side! This gives Uranus a winter that lasts about 21 years!

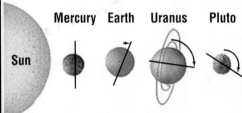

Sun    Mercury    Earth    Uranus    Pluto

## How the Tilts of the Planets Compare

| Planet | Degrees of Tilt |
| --- | --- |
| Mercury | 0 |
| Venus | 177 |
| Earth | 23 |
| Mars | 25 |
| Jupiter | 3 |
| Saturn | 25 |
| Uranus | 98 |
| Neptune | 28 |
| Pluto | 122 |

# How Do Objects Move in the Solar System?

In this lesson, you can . . .

**INVESTIGATE** the ways planets move.

**LEARN ABOUT** our solar system.

**LINK** to math, writing, art, and technology.

# INVESTIGATE

# Planet Movement

**Activity Purpose** Even though you can't feel it, Earth moves through space at nearly 30 kilometers (about 19 mi) per second. At this speed, our planet moves around the sun almost 100 times as fast as most jet planes cruise. You can't make anything move that quickly. So, in this investigation you'll **make a model** that shows how the planets in our solar system move.

## Materials
- index cards
- scissors
- black marker

## Activity Procedure

1. Label one of the cards *Sun*. Label each of the other cards with the name of one of the planets shown in the chart on the next page.

2. Put all of the cards face down on a table and shuffle them. Have each person choose one card.

3. Use the data table to find out which planet is closest to the sun. Continue **analyzing the data** and **ordering** the cards until you have all the planets in the correct order from the sun.

◀ This is Stonehenge, an ancient rock structure located in Britain. Stonehenge may have been used to study and predict the movement of Earth around the sun.

## Planets and Distances from Sun

| Planet | Average Distance from the Sun (millions of km) | Planet | Average Distance from the Sun (millions of km) |
|---|---|---|---|
| Earth | 150 | Pluto | 5900 |
| Jupiter | 778 | Saturn | 1429 |
| Mars | 228 | Uranus | 2871 |
| Mercury | 58 | Venus | 108 |
| Neptune | 4500 | | |

**4** In a gym or outside on a playground, line up in the order you determined in Step 3. (Picture A)

**5** If you have a planet card, slowly turn around as you walk at a normal pace around the sun. Be sure to stay in your own path. Do not cross paths with other planets. After everyone has gone around the sun once, **record** your **observations** of the planets and their movements.

Picture A

## Draw Conclusions

1. The sun and the planets that move around it are called the solar system. What is the order of the planets, starting with the one closest to the sun?

2. What did you **observe** about the motion of the planets?

3. **Scientists at Work** Why did you need to **make a model** to study how planets move around the sun?

**Investigate Further** Look again at the distances listed in the data table. How could you change your model to make it more accurate?

**Process Skill Tip**

Models are used when direct observations aren't possible. The huge size of the solar system is one reason scientists **make models** to study this complex collection of space objects.

# Our Solar System

## The Sun

In the investigation you made a model of our solar system. A solar system is a group of objects in space that move around a central star. Our sun is a star, a burning sphere (SFEER) of gases. This enormous fiery ball is more than 1 million kilometers (about 621,000 mi) in diameter. The sun is the largest object in our solar system. It is larger than the rest of the objects in the solar system put together.

The sun puts out a lot of energy in all directions. In fact, it is the source of almost all the energy in our solar system. Some of this energy reaches Earth as light, and some reaches it as heat.

Two features of the sun's surface are shown on this page. The dark areas, called *sunspots*, are cooler than the rest of the sun's surface and don't give off as much light. The red streams and loops of gases that shoot out from the sun are called *prominences* (PRAHM•ih•nuhn•suhs). These hot fountains often begin near a sunspot. They can be thousands of kilometers high and just as wide. Sunspots and prominences usually last for only a few days. Some can last for a few months.

✔ **What is the largest object in our solar system?**

### FIND OUT

- **about the star we know as the sun**
- **the ways objects move in our solar system**

### VOCABULARY

solar system
star
planet
asteroid
comet
orbit
axis

The sun is the largest object in our solar system. The next largest object, Jupiter, is small compared to the sun. Earth is even smaller. ▼

Jupiter

Earth

Distances not to scale

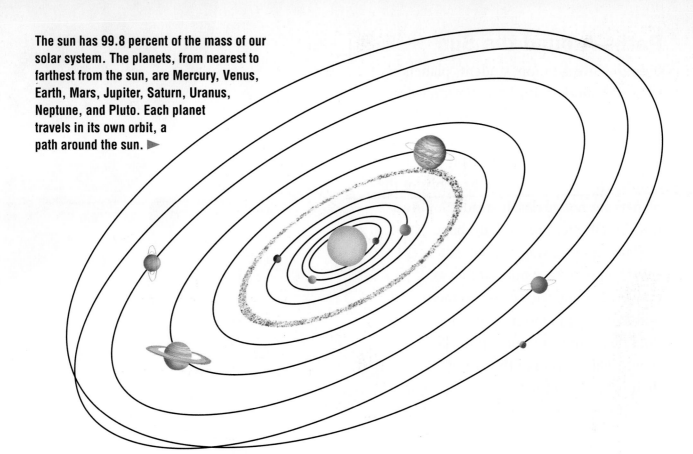

The sun has 99.8 percent of the mass of our solar system. The planets, from nearest to farthest from the sun, are Mercury, Venus, Earth, Mars, Jupiter, Saturn, Uranus, Neptune, and Pluto. Each planet travels in its own orbit, a path around the sun. ▶

## Other Objects in Our Solar System

As you saw in the investigation, our solar system is made up of the sun and nine planets. It also includes moons around the planets, and asteroids and comets.

A **planet**, such as Earth and its eight neighbors, is a large object that moves around a star. Most planets in our solar system also have at least one natural satellite (SA•tuhl•yt), or object that moves around it. These satellites are called *moons*. Earth and Pluto each have only one moon. Jupiter and Saturn, on the other hand, each have many moons.

Asteroids and comets are other objects that move around the sun. **Asteroids** are small and rocky. Most of them are scattered in a large area between the orbit paths of Mars and Jupiter. Some scientists hypothesize that these asteroids are pieces of a

planet that never formed. All the asteroids put together would make an object less than half the size of Earth's moon.

A **comet** is a small mass of dust and ice that orbits the sun in a long, oval-shaped path. When a comet's orbit takes it close to the sun, some of the ice on the comet's surface changes to water vapor and streams out to form a long, glowing tail.

✔ **Name the objects in our solar system.**

◀ As a comet orbits the sun, its tail always points away from the sun. Comet Hale-Bopp passed near Earth in April 1997. Its orbit is so big that it will not be seen from Earth again for 2380 years.

# Paths Around the Sun

In the investigation, student "planets" moved around a "sun." An object *revolves* as it moves around another object. The path of an object as it revolves is called an **orbit**. The time for one complete orbit by a planet around the sun is its *year*. The orbits of the planets are not circles. Instead, they are a little bit elliptical, or oval, in shape.

Have you ever watched ice skaters spin? You may have noticed that as they bring their arms closer to their bodies, they spin faster. When they hold their arms out, they spin more slowly. The motion of the planets in their orbits is a little like the motion of a spinning ice skater. Planets with orbits closer to the sun move faster around the sun than those with orbits that are farther away.

✔ **Where in our solar system are the planets that orbit fastest?**

▲ As Earth *revolves* around the sun, it also spins, or *rotates,* around an imaginary line. This imaginary line, which runs through both poles of a planet, is called an **axis** (AK•sis).

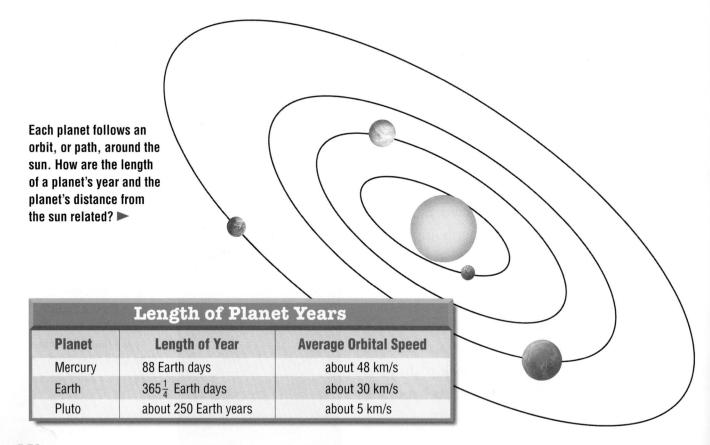

Each planet follows an orbit, or path, around the sun. How are the length of a planet's year and the planet's distance from the sun related? ▶

## Length of Planet Years

| Planet | Length of Year | Average Orbital Speed |
|---|---|---|
| Mercury | 88 Earth days | about 48 km/s |
| Earth | $365\frac{1}{4}$ Earth days | about 30 km/s |
| Pluto | about 250 Earth years | about 5 km/s |

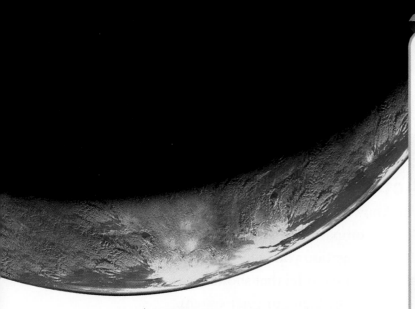

▲ Earth's rotation on its axis causes day and night. The side of Earth that faces the sun has day. At the same time, the opposite side of Earth has night.

## Summary

Our solar system is made up of the sun, nine planets and their moons, asteroids, and comets. Each planet revolves in an elliptical orbit around the sun and rotates on its own axis.

## Review

1. What is the sun?
2. In what way are the planets, comets, and asteroids alike?
3. How does a planet's distance from the sun affect its orbit speed?
4. **Critical Thinking** What might happen if the number of sunspots got much larger?
5. **Test Prep** What provides most of the heat and light to our solar system?
   A the comet Hale-Bopp
   B asteroids
   C the sun
   D Jupiter

# LINKS

### MATH LINK

**The Size of the Sun** The sun's diameter is about 1,392,000 km. About 109 Earths or 9 Jupiters would fit side by side across the sun. Use this information to figure out how many Earths would fit across Jupiter.

### WRITING LINK

**Expressive Writing—Poem** Think about all the things you have learned about our solar system. Write a poem for your family describing how things move in our solar system or what makes up the solar system.

### ART LINK

**Drawing an Ellipse** Tie the ends of a piece of string together to make a loop. Put the loop around two pushpins stuck several inches apart in a thick piece of cardboard. Place the point of a pencil inside the loop so the pencil touches the string. Keep the string tight as you draw around the pushpins. The shape you have drawn is an ellipse. What happens to the shape drawn if you move the pins closer together? Try it.

### TECHNOLOGY LINK

Learn more about how asteroids move around the sun by visiting the National Air and Space Museum Internet site. **www.si.edu/harcourt/science**

 Smithsonian Institution®

# What Are the Planets Like?

In this lesson, you can . . .

**INVESTIGATE** distances between planets.

**LEARN ABOUT** the planets in our solar system.

**LINK** to math, writing, technology, and other areas.

## INVESTIGATE

# Distances Between Planets

**Activity Purpose** If you've ever used a map, you know what a scale model is. A scale model is a way to compare large distances in a smaller space. In this investigation you will **use measurements** to **make a scale model** that shows the distances between planets in our solar system.

### Materials

- piece of string about 4 m long
- meterstick
- 9 different-colored markers

## Activity Procedure

**1** Copy the chart shown on the next page.

**2** At one end of the string, tie three or four knots at the same point to make one large knot. This large knot will stand for the sun in your model.

**3** In the solar system, distances are often measured in astronomical units (AU). One AU equals the average distance from Earth to the sun. In your model, 1 AU will equal 10 cm. Use your meterstick to accurately measure 1 AU from the sun on your model. This point stands for Earth's distance from the sun. Use one of the markers to mark this point on the string. Note in your chart which color you used. (Picture A)

◀ Europa (you•ROH•puh) is one of Jupiter's many moons. This natural satellite has a diameter of 3100 kilometers (about 1925 mi) and takes about $3\frac{1}{2}$ Earth days to orbit Jupiter.

| Planet | Average Distance from the Sun (km) | Average Distance from the Sun (AU) | Scale Distance (cm) | Marker Color | Planet's Diameter (km) |
|--------|-----------|-----------|-----------|-----|-----------|
| Mercury | 58 million | $\frac{4}{10}$ | 4 | | 4876 |
| Venus | 108 million | $\frac{7}{10}$ | 7 | | 12,104 |
| Earth | 150 million | 1 | | | 12,756 |
| Mars | 228 million | 2 | | | 6794 |
| Jupiter | 778 million | 5 | | | 142,984 |
| Saturn | 1429 million | 10 | | | 120,536 |
| Uranus | 2871 million | 19 | | | 51,118 |
| Neptune | 4500 million | 30 | | | 49,532 |
| Pluto | 5900 million | 39 | | | 2274 |

4. Complete the Scale Distance column of the chart. Then measure and mark the position of each planet on the string. Use a different color for each planet, and **record** in your table the colors you used.

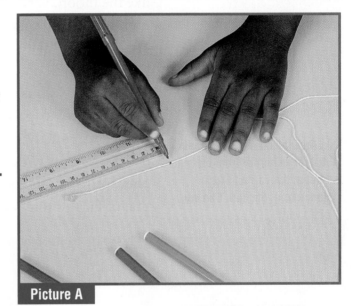

Picture A

## Draw Conclusions

1. In your **model**, how far from the sun is Mercury? How far away is Pluto?

2. What advantages can you think of for using AU to measure distances inside the solar system?

3. **Scientists at Work** Explain how it helped to **make a scale model** instead of trying to show actual distances between planets.

**Investigate Further** You can use a calculator to help make other scale models. The chart gives the actual diameters of the planets. Use this scale: Earth's diameter = 1 cm. Find the scale diameters of the other planets by dividing their actual diameters by Earth's diameter. Make a scale drawing showing the diameter of each planet.

**Process Skill Tip**

**Models** are made to study objects or events that are too small or too large to observe directly. A **scale model** shows large objects or areas in smaller sizes so that they can be more easily studied.

# The Planets

**FIND OUT**

- **about the planets in our solar system**
- **how moons and rings may have formed**

**VOCABULARY**

inner planets
outer planets
gas giants

## The Inner Planets

The area of the asteroid belt can be thought of as a dividing line between two groups of planets, the inner and outer planets. The **inner planets**—Mercury, Venus, Earth, and Mars—lie between the sun and the asteroid belt. Like the asteroids, the inner planets are rocky and dense. Unlike the asteroids, these planets are large and, except for Mercury, have atmospheres.

*Mercury*, the planet closest to the sun, is about the size of Earth's moon. Mercury, which is covered with craters, even looks like the moon. Very small amounts of some gases are present on Mercury, but there aren't enough of them to form an atmosphere.

*Venus*, the second planet from the sun, is about the same size as Earth. But Venus is very different from Earth. Venus is dry and has a thick atmosphere that traps heat. The temperature at the surface is about 475°C (887°F). The thick atmosphere presses down on Venus with a weight 100 times that of Earth's atmosphere. Also, Venus spins on its axis in a direction opposite from that of Earth's rotation.

This drawing shows the planets in the correct order from the sun but not at the correct size or distance from the sun. Can you explain why? ▼

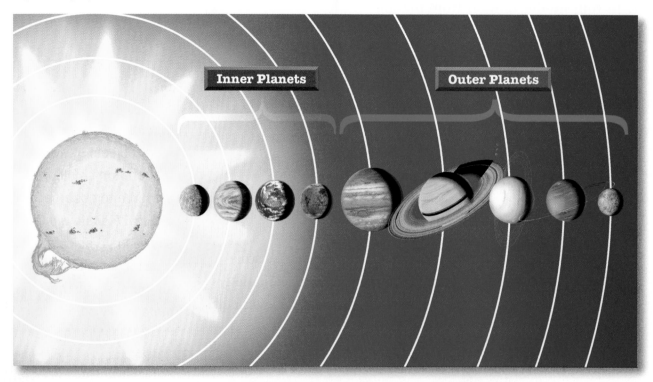

Inner Planets          Outer Planets

Mercury has a diameter of 4876 kilometers (about 3031 mi) and is 58 million kilometers (about 36 million mi) from the sun. This inner planet has no moons and takes about 59 Earth days to make one rotation, or turn once on its axis. Mercury orbits the sun in about 88 Earth days.

*Earth*, the third planet from the sun, is the largest of the inner planets. It has one natural satellite, the moon. Earth is the only planet that has liquid water. It is also the only known planet that supports life. Earth's atmosphere absorbs and reflects the right amount of solar energy to keep the planet at the correct temperature for living things such as humans to survive.

*Mars*, the fourth planet from the sun, is sometimes called the Red Planet because its soil is a dark reddish brown. Mars has two moons and the largest volcano in the solar system—Olympus Mons (oh•LIHM•puhs MAHNS). Space probes have shown us that nothing lives on Mars. Dust storms can last for months and affect the whole planet. Although no liquid water has been found on Mars, it is believed that liquid water once existed there. This is because probes and satellites have found deep valleys and sedimentary rocks. These features probably were formed by flowing water.

✔ **List the inner planets in order from the sun.**

Venus has a diameter of 12,104 kilometers (about 7517 mi) and is 108 million kilometers (about 67 million mi) from the sun. This inner planet has no natural satellites. It takes Venus about 243 Earth days to make one rotation and 225 Earth days to orbit the sun.

Earth has a diameter of 12,756 kilometers (about 7922 mi). Our planet is 150 million kilometers (about 93 million mi) from the sun. Earth has one moon and takes almost 24 hours to complete one rotation on its axis. It takes about 365 days to orbit the sun.

Not to scale

Mars has a diameter of 6794 kilometers (about 4230 mi) and is 228 million kilometers (about 142 million mi) from the sun. This inner planet has two moons. Mars completes a rotation in about 24.5 hours. Mars takes about 687 Earth days to complete one orbit around the sun.

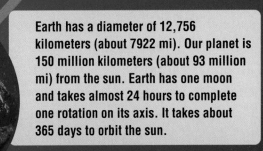

Jupiter has a diameter of 142,984 kilometers (about 88,700 mi) and is 778 million kilometers (about 483 million mi) from the sun. This outer planet takes about 10 hours to make one rotation and almost 12 Earth years to orbit the sun.

Saturn is a gas giant that has a diameter of 120,536 kilometers (about 74,900 mi). Saturn is 1429 million kilometers (about 887 million mi) from the sun. This outer planet takes about 10 hours to turn once on its axis. Saturn orbits the sun in a little more than 29 Earth years.

## The Outer Planets

On the other side of the asteroid belt are the **outer planets**—Jupiter, Saturn, Uranus, Neptune, and Pluto. Four of these planets— Jupiter, Saturn, Uranus, and Neptune—are large spheres made up mostly of gases. Because of this, these planets are often called the **gas giants**.

*Jupiter* is the largest planet in our solar system. A thin ring that is hard to see surrounds it. At least 16 moons orbit around it. Jupiter's atmosphere is very active. Its energy causes a circular storm known as the Great Red Spot. This weather, which is a lot like a hurricane, has lasted more than 300 years. It is so big around that three Earths would fit inside it.

*Saturn* is a gas giant known for its rings. Space probes have found that other planets also have rings. But Saturn's are so wide and so bright that they can be seen from Earth through a small telescope. Saturn has at least 18 named moons.

Not to scale

Uranus (YOOR•uh•nuhs), the seventh planet from the sun, is the most distant planet you can see without using a telescope. Uranus, a blue-green ball of gas and liquid, has at least 15 moons as well as faint rings around it.

Neptune, the farthest away of the gas giants, has at least eight moons and a faint ring. It also has circular storms, but none have lasted as long as the Great Red Spot on Jupiter.

In the investigation you saw that Pluto is the planet farthest from the sun. If you completed the Investigate Further, you also learned that Pluto is the smallest planet. From Pluto's surface the sun looks like a very bright star. Little heat or light reaches Pluto or its one moon. Unlike the other outer planets, Pluto is not a gas giant. Instead, Pluto has a rocky surface that is probably covered by frozen gases.

✓ **Which planets are gas giants?**

Uranus has a diameter of 51,118 kilometers (about 31,700 mi). This planet is 2870 million kilometers (1782 million mi) from the sun. Uranus makes one rotation in 17 hours and one orbit around the sun in about 84 Earth years.

Pluto has a diameter of 2274 kilometers (about 1366 mi) and is 5900 million kilometers (about 3664 million mi) from the sun. It takes Pluto about 7 days to complete one rotation and 249 Earth years to complete one revolution.

Neptune has a diameter of 49,532 kilometers (about 30,740 mi) and is 4500 million kilometers (about 2795 million mi) from the sun. Neptune completes one rotation in a little more than 16 hours and one revolution in about 165 Earth years.

▲ Titan, Saturn's largest moon, has a diameter of 5150 kilometers (about 3200 mi). Titan has no clouds in its atmosphere and is very cold.

▲ The rings around Saturn are made up of dust, ice crystals, and small bits of rock coated with frozen water. The rings are about 270,000 kilometers (about 167,700 mi) across but only about 10 kilometers (about 6 mi) thick.

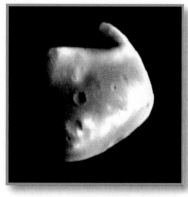

Io (EYE•oh) is a moon of Jupiter. Io has a diameter of 3630 kilometers (about 2254 mi) and has several active volcanoes on its surface. The volcanoes are the big orange patches. ▶

▲ Deimos (DY•muhs) is one of the two Martian moons. It has many craters and an uneven shape. It is about 14 kilometers (9 mi) in diameter.

## Moons and Rings

Every planet except Mercury and Venus has at least one natural satellite, or *moon*. Earth's moon is round and rocky, and it has many craters. Others, like the two moons of Mars or the outer moons of Jupiter, are small and rocky and have uneven shapes. Jupiter and Mars orbit near the asteroid belt. So, their moons may be asteroids pulled in by the planets' gravity. The large moons of Jupiter and Saturn are almost like small planets. Io, one of Jupiter's larger moons, has active volcanoes. Titan (TYT•uhn), one of Saturn's moons, has a dense atmosphere that glows red-orange.

Besides having moons, each of the gas giants has a system of rings. These rings are made of tiny bits of dust, ice crystals, and small pieces of rock. Saturn's rings may have formed as a moon was pulled apart by gravity because it got too close to the planet.

✔ **What are planet rings made of?**

▲ Phobos (FOH·buhs) is the other moon that orbits Mars. Like Deimos, it is a small, rocky object. Its diameter is about 22 kilometers (14 mi). Phobos makes three trips around its planet each Martian day.

## Summary

The inner planets—Mercury, Venus, Earth, and Mars—are small and rocky. Four of the five outer planets are gas giants. They are Jupiter, Saturn, Uranus, and Neptune. The outer planet that is farthest from the sun is Pluto, another rocky planet. Most of the planets have at least one moon. The gas giants also have rings.

## Review

1. Name the inner planets, starting with the planet closest to the sun.
2. What can be thought of as the dividing line between the inner planets and the outer planets?
3. Which planets have no moons?
4. **Critical Thinking** Compare and contrast Venus and Earth.
5. **Test Prep** The outer planet that is **NOT** a gas giant is —
   A Jupiter
   B Saturn
   C Neptune
   D Pluto

# LINKS

## MATH LINK

**Graphing Planet Data** Make a bar graph showing the diameter of each planet. Use data from the table on page D61 and a computer program such as *Graph Links*.

## WRITING LINK

**Persuasive Writing—Opinion** Suppose you are a real-estate agent trying to get adults to move to the planet of your choice. Write a newspaper ad pointing out all the benefits of living on your chosen planet.

## ART LINK

**View of a Planet** Paint or draw a realistic landscape of the surface of another planet. Or paint the view of the planet as it would be seen from one of its moons.

## LITERATURE LINK

***The Wonderful Flight to the Mushroom Planet*** Read this book by Eleanor Cameron to find out what happens when two boys go on an adventure in space. Compare the planet with Earth.

## TECHNOLOGY LINK

Learn more about planets and other objects in space by visiting this Internet site.
**www.scilinks.org/harcourt**

SCI LINKS™
THE WORLD'S A CLICK AWAY

# How Do People Study the Solar System?

In this lesson, you can . . .

 **INVESTIGATE** how to make a telescope.

 **LEARN ABOUT** how people study objects in space.

 **LINK** to math, writing, social studies, and technology.

## INVESTIGATE

# Telescopes

**Activity Purpose** Have you ever looked up into the sky at night and wished you could see some of the objects more clearly? Because distances in space are so great, scientists need to use instruments to study what is beyond Earth's atmosphere. In this investigation you will make a simple telescope and use it to **observe** some objects in space.

### Materials

- small piece of modeling clay
- 1 thin (eyepiece) lens
- small-diameter cardboard tube
- 1 thick (objective) lens
- large-diameter cardboard tube

**CAUTION**

## Activity Procedure

1 Press small pieces of clay to the outside of the thin lens. Then put the lens in one end of the small tube. Use enough clay to hold the lens in place, keeping the lens as straight as possible. Be careful not to smear the middle of the lens with clay. (Picture A)

2 Repeat Step 1 using the thick lens and large tube.

◄ This vehicle is the moon rover. Astronauts used it to travel over the surface of Earth's only natural satellite, the moon.

Picture A

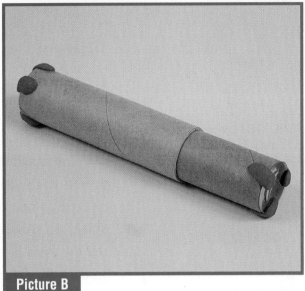

Picture B

**3** Slide the open end of the small tube into the larger tube. You have just made a telescope. (Picture B)

**4** Hold your telescope up, and look through one lens. Then turn the telescope around, and look through the other lens. **CAUTION** **Never look directly at the sun.** Slide the small tube in and out of the large tube until what you see is in focus, or not blurry. How do objects appear through each lens? **Record** your **observations.**

## Draw Conclusions

1. What did you **observe** as you looked through each lens?

2. Using your observations, **infer** which lens you should look through to **observe** the stars. Explain your answer.

3. **Scientists at Work** Astronomers (uh•STRAWN•uh•merz) are scientists who study objects in space. Some astronomers use large telescopes with many parts to **observe** objects in space. How would your telescope make observing objects in the night sky easier?

**Investigate Further** Use your telescope to observe the moon at night. Make a list of the details you can see using your telescope that you can't see using only your eyes.

**Process Skill Tip**

When you **observe** an object, you use your senses to notice details about it. Using an instrument such as a telescope helps you observe objects that are too far away to be seen clearly using only your eyes.

# Space Exploration

## Telescopes

Using nothing more than your eyes, you can see most of the planets in the solar system. But what if you want to see them as more than just points of light in the sky? What if you want to see objects in space that are even farther away than the visible planets? Or what if you want to see smaller objects such as moons? To do any of these things, you need to use a telescope. A **telescope** is a device people use to observe distant objects.

Two very different types of telescopes are used to observe objects in space: *radio telescopes* and *optical telescopes,* or telescopes that use light. There are also two types of optical telescopes. A refracting telescope, which is what you made in the investigation, uses lenses to magnify an object, or make it appear larger. A reflecting telescope uses a curved mirror to magnify an object. Most large telescopes used today are reflecting telescopes.

**FIND OUT**

- **about telescopes**
- **about missions into space**

**VOCABULARY**

**telescope**
**space probe**

Even without using a telescope, you can see dark areas, bright areas, and craters on the moon. ▼

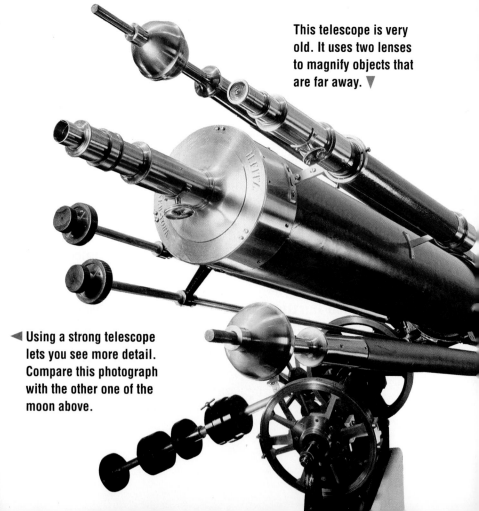

This telescope is very old. It uses two lenses to magnify objects that are far away. ▼

◄ Using a strong telescope lets you see more detail. Compare this photograph with the other one of the moon above.

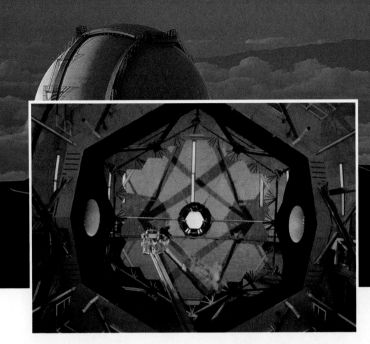

▲ The large mirror of the Keck telescope is made up of many smaller mirrors. They work together to gather light and magnify images of objects in space.

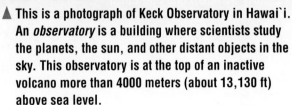

▲ This is a photograph of Keck Observatory in Hawai`i. An *observatory* is a building where scientists study the planets, the sun, and other distant objects in the sky. This observatory is at the top of an inactive volcano more than 4000 meters (about 13,130 ft) above sea level.

Telescopes that scientists use are much larger than the one you made in the investigation. Besides being large, telescopes that are used to study space objects are powerful. Many of them have cameras that constantly take pictures of space. Computers keep the telescopes pointed at the same place in the sky. This allows the powerful lenses and mirrors to collect more light so they can produce brighter and clearer pictures. Astronomers then study the pictures to find out about space objects.

Earth's atmosphere limits what optical telescopes can "see." Moving air causes the "twinkling" of stars. It also blurs pictures taken using optical telescopes. This is why observatories that use optical telescopes are often located high on mountains.

A radio telescope collects radio waves with a large, bowl-shaped antenna. Scientists study images formed by these waves to learn about the objects that gave them off. This radio telescope is in Arecibo (ar•uh•SEE•boh), Puerto Rico. ▶

Scientists have found that stars and other objects in space give off more than just light energy that we can see. Radio telescopes work the way optical telescopes do. But instead of collecting and focusing light, they collect and focus invisible radio waves. Moving air, clouds, and poor weather don't affect radio waves. Computers process the data collected by radio telescopes. The computers then make "pictures" that astronomers can study.

The telescopes you have seen so far in this chapter are Earth-based, or located on Earth's surface. Scientists have also built telescopes for use in space. These telescopes don't have any problems caused by the atmosphere. The Inside Story shows the parts of the most famous space-based telescope, the Hubble Space Telescope.

✓ **What is a telescope?**

## THE INSIDE STORY

# HUBBLE SPACE TELESCOPE

The Hubble Space Telescope, or HST, is a reflecting telescope. Its mirror, which has a diameter of 240 cm (about 94 in.), can "see" details ten times as clearly as telescopes on Earth.

The HST uses sunlight as its energy source. The instrument's solar panels change sunlight into electricity. Each panel is about $2\frac{1}{2}$ meters (about 8 ft) wide and 13 meters (about 42 ft) long.

The main cover tube of the HST protects the telescope as well as other instruments used to study space.

The main mirror of the HST is located near the back of the main cover.

*Mir* was a Russian *space station* in orbit above Earth. Aboard *Mir*, astronauts and scientists from many countries worked for months at a time. They experimented to find out how conditions in space affect things and people. ▼

▲ During some of the Apollo missions, astronauts explored our moon's surface. This spacecraft allowed astronauts to land on the moon. Later they could blast off and return to the main part of their ship, which was circling the moon.

## Crewed Missions

Another way to learn about space is to actually go there. Trips that people take into space are called *crewed missions.* Crewed missions are useful because people can actually find out for themselves what it is like to live and work in space.

The first person to be sent into space was a Russian, Yuri Gagarin, in 1961. Since then, astronauts from many countries have made trips into space. One of the most famous missions was carried out on *Apollo 11,* which was launched by the United States on July 16, 1969. On this mission Neil Armstrong and Edwin "Buzz" Aldrin spent two hours exploring the moon's surface. The United States sent five more crewed missions to the moon before the Apollo program ended in 1972.

Today the United States uses the space shuttle to carry crews, materials, and satellites to and from space. Astronauts on the shuttles do experiments, launch and get back satellites, and repair instruments.

✔ **What is a crewed space mission?**

Astronauts aboard the U.S. space shuttles also do experiments. These astronauts are working in Space Lab, which rides in the cargo area of a space shuttle. ▼

D73

# Space Probes

We have learned much of what we know about the solar system by using space probes. **Space probes** are vehicles that carry cameras, instruments, and other tools. Probes are sent to explore places that are too dangerous or too far away for people to visit. Probes gather data and send it back to Earth for study. The pictures of Mars and Callisto were sent to Earth by probes millions of kilometers away in space.

Some probes have fly-by missions. That is, they fly by the object to be studied but do not land. As they pass the object, they gather data, including pictures. Other probes land on the surfaces of planets. These probes take pictures, collect and analyze rock samples, test for the presence of substances such as water, and collect other data.

✔ **What is a space probe?**

▲ Information about this moon of Jupiter, Callisto, was gathered as the space probe *Galileo* flew past it.

*Sojourner* is a probe that was used to study the surface of Mars. *Sojourner* took thousands of photos of the surface as it gathered information about the Red Planet. ▼

▲ The space probe *Galileo* was sent to Jupiter in 1989. Instruments launched from *Galileo* measured the sizes of particles that make up the clouds of Jupiter and the amounts of hydrogen and helium in Jupiter's atmosphere.

▲ *Voyager 2*, launched in 1977, flew by Jupiter in 1979, Saturn in 1981, and Uranus in 1986. The probe then headed on to Neptune.

## Summary

A telescope is a device people use to observe distant objects. A refracting telescope uses lenses, and a reflecting telescope uses mirrors to magnify an object. Radio telescopes collect and focus radio waves. The Hubble Space Telescope is an optical telescope in orbit around Earth. Crewed missions and space probes are other ways to study objects in space.

## Review

1. What are two types of telescopes?
2. What limits the things an Earth-based optical telescope can "see"?
3. What kind of work do astronauts on a crewed space mission do?
4. **Critical Thinking** What are the advantages of sending crewed missions instead of probes into space? The disadvantages?
5. **Test Prep** An instrument that uses lenses to magnify distant objects is a —
   A radio telescope
   B refracting telescope
   C reflecting telescope
   D space station

# LINKS

## MATH LINK

**Faster than a Speeding Rocket** When Earth and Jupiter are closest together, they are about 630 million km apart. *Voyager 2* took two years to reach Jupiter. About how far did it travel each year?

## WRITING LINK

**Informative Writing—Description** Suppose you are an astronaut who will take part in the next space-shuttle mission. Find out about a typical day on the space shuttle. Then write a composition for a friend describing a typical day of your mission.

## SOCIAL STUDIES LINK

**International Space Station** Find out which countries are building parts of the International Space Station. Locate each country on a map or globe. Make a chart that lists the name of each country, its continent, and which parts of the space station it is building.

## TECHNOLOGY LINK

To learn more about pictures from space watch *Hubble Images* on the **Harcourt Science Newsroom Video.**

# What Are Constellations?

In this lesson, you can . . .

**INVESTIGATE** how to make a constellation box.

**LEARN ABOUT** star patterns called constellations.

**LINK** to math, writing, literature, and technology.

◀ This instrument is called a *sextant* (SEKS•tuhnt). It was commonly used by sailors hundreds of years ago. By measuring the position of the sun above the horizon, you can find out your exact location on the ocean.

# INVESTIGATE

# Constellations

**Activity Purpose**  On a clear night, away from city lights, the sky is filled with twinkling stars. Some of these stars form patterns in the sky. Ancient people named these patterns after the objects or persons the patterns reminded them of. In this investigation you will **make a model** to show a pattern made by stars in the sky.

## Materials

- safety goggles
- scissors
- shoe box with lid
- penlight
- black construction paper
- straight pin
- thin, wooden skewer
- masking tape

CAUTION

## Activity Procedure

1 **CAUTION** **Put on your safety goggles.** Use the scissors to carefully cut a rectangle about 10 cm by 6 cm from one end of the box.

2 At the other end of the box, trace the diameter of the penlight's head. Use the scissors to carefully cut out the circle.

3 Cut two or three rectangles from the black construction paper. Make sure they are a little larger than the rectangular opening in the shoe box.

4 Use the pin and the wooden skewer to carefully make small holes in the construction paper. Make a different pattern on each rectangle. (Picture A)

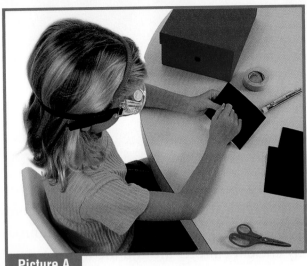

Picture A

Picture B

**5** Put the head of the penlight into the circular opening in the shoe box. Hold the penlight in place with masking tape. Put the lid on the box.

**6** Put one of your rectangles over the opening at the other end of the box. Tape it in place. (Picture B)

**7** Darken the room. Hold the box about 0.5 m from a white wall, and turn on the penlight. **Record** your **observations** of your star pattern and those of your classmates.

**8** Work with a partner to find a pattern of stars that looks like an object or person. Name your pattern, and make up a story about it.

## Draw Conclusions

1. What did the star patterns look like?

2. Could you find a group of "stars" that looked like the shape of an object or person? These patterns found among the stars are called constellations. Describe your constellation.

3. **Scientists at Work** Not all stars are the same distance from Earth. Also, not all stars shine with the same brightness. Think about your **model**. What do you think the larger holes stand for? What do the smaller holes stand for?

**Investigate Further** Again, project your star pattern on the wall, this time with the lights on. How does what you see in the light **compare** with what you saw in the dark? **Infer** why you can't see most stars during the day.

**Process Skill Tip**

Scientists often use models to study objects or events that they can't directly observe. **Making models** of space objects makes it much easier to study the objects and the relationships among them.

# Constellations

## Patterns of Stars

**FIND OUT**

• about constellations

• how constellations seem to change their positions during the year

**VOCABULARY**

constellation

If you look into the night sky, you'll notice that many of the brightest stars form patterns much like those you made and projected on the wall in the investigation. Each pattern or group of stars in the sky is called a **constellation** (kahn•stuh•LAY•shuhn).

The people of many ancient cultures around the world saw constellations as outlines of objects, mythological (MITH•uh•lahj•ih•kuhl) characters, or animals. Those people made up stories to explain how the object, character, or animal came to be in the night sky.

Orion, one of the brightest constellations, is an example of a mythological character. In ancient Greek myths, Orion was a hunter who annoyed the gods by bragging all the time. When they got tired of him, the gods sent a scorpion to bite and kill him. Afterward, they felt sorry for Orion and placed him in the sky.

✔ **What is a constellation?**

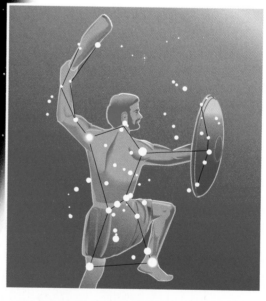

◀ If you think about the legend, you can see Orion in this group of stars.

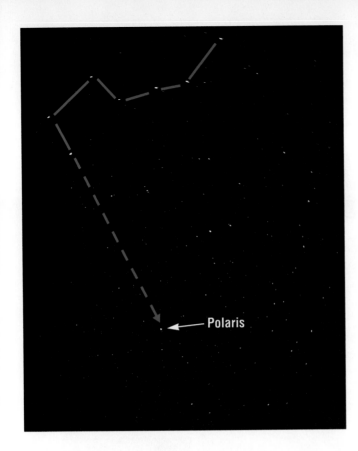

Polaris

◀ The Big Dipper is an easy constellation to find. Then you can use it to find Polaris. Use the "pointer stars" in the end of the dipper's bowl, and follow them to Polaris. Polaris marks the end of the Little Dipper's handle.

## Stars as Tools for Navigation

Because Earth rotates on its axis, most constellations appear to rise in the east and set in the west during the night. As Earth revolves around the sun, most constellations seem to appear in different positions in the sky. However, if you look toward the northern sky, you will see a group of stars that are visible all night, all year long. Instead of appearing to rise and set, these stars seem to circle Earth's North Pole each night. They are called circumpolar (sir•kuhm•POH•ler) stars. Polaris (poh•LAR•ihs), the North Star, is located almost directly above the North Pole. This star appears in the same place all night long, every night of the year.

If you are in the Northern Hemisphere and can find Polaris in the night sky, you can find which direction is north. And if you know where north is, you can find your way when you are traveling. Sailors used stars such as Polaris to navigate, or find their way on the ocean. Besides knowing which direction is north, sailors can look at other stars and tell how far east or west they have traveled.

It's not as easy to find your way if you are in the Southern Hemisphere. There is no star above the South Pole. Instead, travelers look for the smallest constellation in the sky—the Southern Cross. The long side of this circumpolar constellation of four bright stars and several dimmer ones points toward the South Pole.

✔ **Why is Polaris called the North Star?**

The Southern Cross is used to locate south in the Southern Hemisphere. It is not directly over the South Pole, but it is near it. There is no bright star called the "South" Star. ▶

## Stars as Calendars

As Earth revolves around the sun, the constellations that rise and set appear to rise a little earlier in the east each evening. This means that as the seasons change, we see different constellations in the night sky. For example, in the Northern Hemisphere, Orion is high in the night sky during the winter, while the constellation Scorpio is visible only in summer.

Ancient people used these seasonal changes of star patterns as calendars. Some historians think that ancient people watched constellations to decide when to plant and harvest their crops. Seeing Leo and Virgo in the night sky, for example, may have told them that the last frosts of the year had happened and that it was now safe to plant. This is not very different from a farmer today using a paper calendar to count days.

✔ **How did ancient people use the seasonal appearance of certain constellations?**

▲ Although constellations near the poles are always visible and never rise or set, their positions do change with the seasons. This is the North Polar sky in summer.

▲ This is the same section of sky as above at the same time of night, but in winter. All the stars and constellations are still here, but their positions appear to have rotated.

This half of a star chart shows the positions of the constellations in the sky from February to August. To use a chart like this one, face north. Hold the chart so the current month is at the top. Then once you find the Big Dipper and Polaris, it is easy to find other constellations. ▶

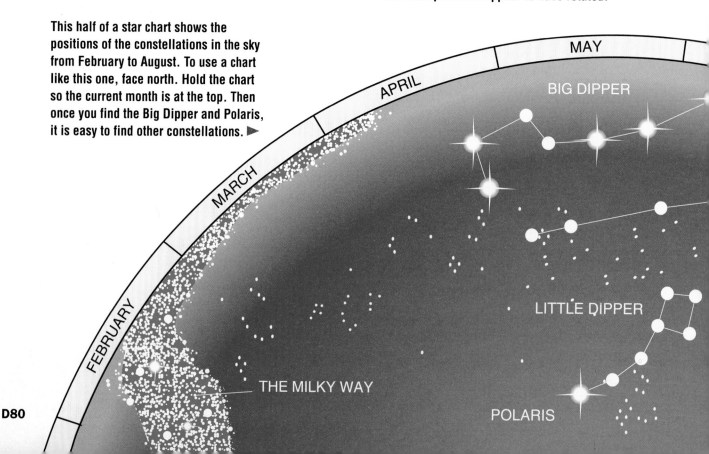

MAY

APRIL

BIG DIPPER

MARCH

LITTLE DIPPER

FEBRUARY

THE MILKY WAY

POLARIS

## Summary

In the sky, stars form patterns called constellations. Circumpolar constellations appear to circle the poles. Stars and constellations can be used for navigation. Ancient people may have used seasonal constellations as calendars.

## Review

1. What is a constellation?
2. How do circumpolar stars differ from other stars?
3. Describe how to locate the North Star.
4. **Critical Thinking** If you were lost in Australia on a clear night, explain how you could find your way.
5. **Test Prep** A constellation that is seen high in the night sky at only certain times of the year is —

    A circumpolar
    B seasonal
    C mythical
    D navigational

# LINKS

 **MATH LINK**

**Distances in Space** Find out about two units used to measure distances in space: the astronomical unit, or AU, and the light-year. Tell when each of these units of measurement is used.

 **WRITING LINK**

**Informative Writing—Narration** Choose a constellation, and find out the story behind it. Write a story for a younger child explaining a myth that ancient people told about your constellation.

 **LITERATURE LINK**

***The Ultimate Guide to the Sky*** Go to your library, and check out this book by John Mosley. On a clear night, work with an adult family member to locate and identify some of the constellations described in this book. Note that star viewing is best done away from city lights and between 9 P.M. and 10 P.M.

 **TECHNOLOGY LINK**

Visit the Harcourt Learning Site for related links, activities, and resources.
**www.harcourtschool.com**

WELCOME TO
THE
LEARNING
SITE

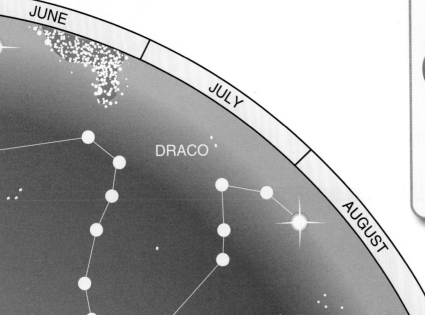

JUNE

JULY

DRACO

AUGUST

# Discovering the Planets

Five of the nine planets in our solar system can be seen using just your eyes. These five—Mercury, Venus, Mars, Jupiter, and Saturn—were all known to astronomers before 1700. These early astronomers named the planets for gods and goddesses of Roman mythology.

## Galileo and the Telescope

The invention of the telescope made astronomy a more complex science. Galileo, an Italian scientist and inventor who lived from 1564 to 1642, was the first person to use a telescope to view the stars and planets. Compared to telescopes now, his telescope wasn't very powerful. It magnified, or enlarged the view, only about 20 or 30 times. However, using just that weak telescope, Galileo was the first person to observe mountains on the moon, sunspots, and the phases of Venus. In 1610 Galileo was the first to see moons around Jupiter. His observations of planets and moons showed that not all bodies in space circled Earth, as many people at that time believed.

◀ These ink drawings are how Galileo recorded his telescope observations of Earth's moon.

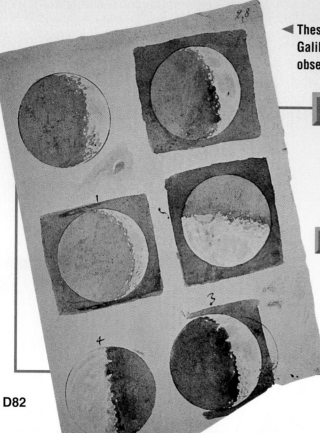

## The History of Planet Discoveries

Uranus 1781—
**Uranus discovered**

| 1600 A.D. | 1700 A.D. |
|---|---|

Jupiter 1610
**Jupiter's moons discovered**

## Discovering Uranus

Other scientists slowly improved on Galileo's telescope design. William Herschel, a British astronomer, discovered Uranus, the seventh planet from the sun. He used a telescope that his sister, Caroline, helped him build. On the night of March 13, 1781, Herschel saw something that he knew was not a star. He first thought it was a comet. Over time he mapped the orbit of the object and realized it was a planet, which was later named Uranus. For a long time, others had a hard time seeing what Herschel had seen, because their telescopes weren't as good.

## Using Math to Find New Planets

Two astronomers, the English John C. Adams and the French Urbain Leverrier (oor•BAN luh•vair•YAY), found the location of the eighth planet at about the same time. They did not see the planet themselves. Using mathematics, they predicted its location. Then they sent their predictions to other scientists.

Adams sent his prediction to the Astronomer Royal of England, who paid little attention to it. Leverrier sent his prediction to the Urania (oo•RAHN•ee•uh) Observatory in Berlin, Germany. The director there, Johann Galle (YOH•hahn GAH•luh), and his assistant used Leverrier's research and found Neptune on September 23, 1846.

Pluto was also discovered through mathematics. In 1905 American astronomer Percival Lowell noticed that something seemed to be affecting the orbits of Uranus and Neptune. He hypothesized that a ninth planet was the cause. He searched unsuccessfully for it until his death in 1916.

In 1930 Pluto appeared as a small dot on three photographs Clyde Tombaugh took at Lowell Observatory, which was built by Percival Lowell.

## Think About It

1. How have improvements in technology helped astronomers?
2. How did scientists Urbain Leverrier and Johann Galle work together?

Ganymede, one of Jupiter's moons

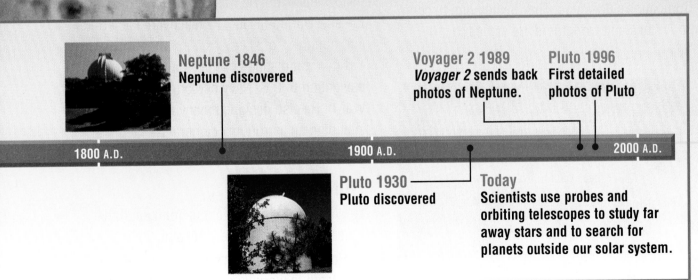

Neptune 1846
**Neptune discovered**

Voyager 2 1989
*Voyager 2* sends back photos of Neptune.

Pluto 1996
**First detailed photos of Pluto**

1800 A.D.        1900 A.D.                    2000 A.D.

Pluto 1930
**Pluto discovered**

**Today**
Scientists use probes and orbiting telescopes to study far away stars and to search for planets outside our solar system.

# Clyde Tombaugh
## ASTRONOMER, INVENTOR

*"I think the driving thing was curiosity about the universe. That fascinated me. I didn't think anything about being famous or anything like that, I was just interested in the concepts involved."*

Tombaugh in 1995 with his first telescope

**W**orking on a farm in Kansas taught Clyde Tombaugh to be persistent and creative with materials at hand. He was always interested in astronomy. His uncle and father had a telescope, which they gave him when he was 9 years old. By the time Tombaugh was 20, he decided to build his own telescope. He used part of a dairy machine for the base and part of his father's 1910 Buick.

Later, Tombaugh's uncle asked Tombaugh to build a telescope for him. Tombaugh also built a better telescope for him. With that homemade telescope, he observed Mars and Jupiter. He drew what he saw and sent his sketches to the Lowell Observatory in Flagstaff, Arizona. The scientists at the observatory were impressed by Tombaugh's drawings. They invited him to come to Arizona and work there. He stayed for 14 years.

On February 18, 1930, Tombaugh discovered the planet Pluto. Other scientists had predicted its existence, but he was the first to locate it in the sky. During his life, he discovered asteroids and hundreds of stars.

After completing college, Tombaugh taught navigation to U.S. Navy personnel during World War II. He also designed many new instruments, including an astronomy camera. He taught at New Mexico State University for nearly 20 years.

## THINK ABOUT IT

1. What have you read about Tombaugh that leads you to think he is resourceful?

2. How was Tombaugh's discovery of Pluto related to work by other scientists?

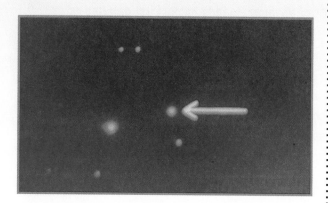

Telescope photo of Pluto, 1930

## Sundial

### *How can you make an instrument that uses the sun to tell time?*

**Materials**
- small ball of clay
- short pencil
- cardboard, about 15 cm x 20 cm

**Procedure**

1 Use the lump of clay to stand the short pencil in the center of the cardboard. Make sure the sharpened end of the pencil is up.

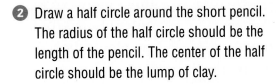

2 Draw a half circle around the short pencil. The radius of the half circle should be the length of the pencil. The center of the half circle should be the lump of clay.

3 Put your sundial on a windowsill that gets sun all day. Each hour, trace the shadow of the pencil on the cardboard. Mark the time at the end of each line. Do this for six hours.

4 On the next sunny day, use your sundial to tell time.

**Draw Conclusions**

How does your sundial use Earth's movement to tell time?

## Moving Constellations

### *How do the positions of the constellations change?*

**Materials**
- ruler

**Procedure**

1 Observe the Big Dipper and the Little Dipper twice each night. One time should be in the early evening. The second time should be at least one hour later.

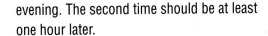

2 Find something to use as a reference, such as the top of a fence or tree. Hold the ruler at arm's length, and measure the distance from the reference point to the constellations. Record your observations as drawings.

3 Repeat Steps 1 and 2 at the same time each night for four weeks. Remember to put the dates on your drawings.

**Draw Conclusions**

How did the constellations change their positions? What caused the changes?

## Vocabulary Review

Use the terms below to complete the sentences. The page numbers in ( ) tell you where to look in the chapter if you need help.

**solar system** (D56)     **inner planets** (D62)
**star** (D56)     **outer planets** (D64)
**planet** (D57)     **gas giants** (D64)
**asteroid** (D57)     **telescope** (D70)
**comet** (D57)     **space probe** (D74)
**orbit** (D58)     **constellation** (D78)
**axis** (D58)

1. A ____ is a group of planets and their moons that orbit a central star.

2. Venus is a ____ that revolves around the sun.

3. An ____ is an imaginary line around which a planet rotates, or spins.

4. A group of stars that forms a pattern in the night sky is called a ____.

5. The four planets nearest the sun are called the ____.

6. An ____ is a rocky object that orbits the sun in a path between Mars and Jupiter.

7. A ____ is a vehicle sent into space in order to explore places too dangerous or too far away for people to visit.

8. The path a planet takes around the sun is its ____.

9. A ____ is a burning ball of gases.

10. Neptune is one of the four ____.

11. A ____ is an instrument used to observe distant objects.

12. A ____ is a space object made of ice, dust, and gases.

13. The five planets on the outer side of the asteroid belt are called the ____.

## Connect Concepts

List the planets, and classify them as to their position in the solar system. Be sure to list them in the correct order from the sun.

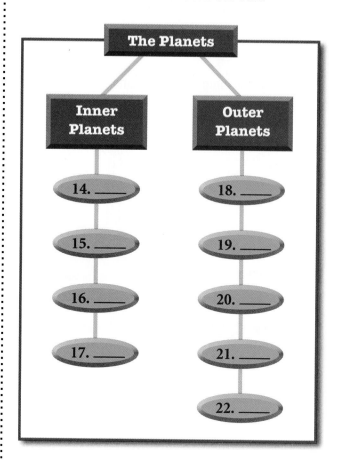

## Check Understanding

Write the letter of the best choice.

23. The sun provides ____ of the energy to the solar system.
    **A** most          **C** none
    **B** little          **D** some

24. An imaginary line that runs through both poles of a planet is its —
    **F** star          **H** comet
    **G** orbit          **J** axis

25. ____ is the outer planet that is **NOT** a gas giant.
    A Saturn    C Uranus
    B Pluto     D Jupiter

26. ____ has a ring.
    F Jupiter    H Earth
    G Pluto      J Venus

27. This type of telescope never has problems seeing through Earth's atmosphere. It uses lenses to enlarge an object.
    A Earth-based telescope
    B reflecting telescope
    C refracting telescope
    D space-based telescope

28. Trips that people take into space are called —
    F crewed missions
    G uncrewed missions
    H space probes
    J observatories

29. Constellations that appear to circle the poles are called ____ constellations.
    A seasonal      C planet
    B circumpolar   D star

## Critical Thinking

30. Why do you think some stars in constellations look brighter than others?

31. Most asteroids are in the asteroid belt between the orbits of Mars and Jupiter. But some asteroids have "escaped" and have long, oval-shaped orbits like those of the comets. How do you think these asteroids escaped the asteroid belt?

## Process Skills Review

32. In Lesson 1 you **made a model** showing how planets rotate on their axes and revolve around the sun. How would you make a model to show the motions of Jupiter and its many moons?

33. In Lesson 2 you **made a model** to show the distances between the planets. What activity of planets would be modeled by cars racing around a racetrack?

34. In Lesson 3 you made a telescope to **observe** the night sky. Why did you need a telescope for this? What would you see if you didn't use a telescope?

35. In Lesson 4 you **made a model** of star patterns. Do you think you could model actual constellations by using this method? How might this kind of model be useful?

## Performance Assessment

### Model Solar System

With a partner, draw a model of an imaginary solar system. Include one star, five planets, and two comets. Also include at least one more object that would be found in a solar system. Compare your model solar system with the one we live in.

# Unit Project Wrap Up

## Here are some ideas for ways to wrap up your unit project.

### Produce a Weather Program

Videotape your daily weather reports and forecasts, and play the tapes for the school.

### Write Guidelines

Write directions that tell how to use your weather instruments. Include information about why and how each instrument works.

### Make a Chart

For a week, evaluate how accurately the weather was forecast by professional meteorologists. Make a chart to organize your data.

### Investigate Further

How could you make your project better? What other questions do you have? Plan ways to find answers to your questions. Use the Science Handbook on pages R2-R9 for help.

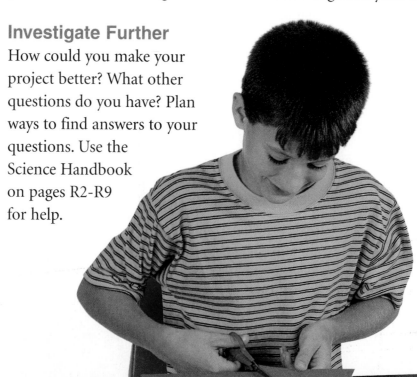

# References

## Science Handbook

# Planning an Investigation

When scientists observe something they want to study, they use the method of scientific inquiry to plan and conduct their study. They use science process skills as tools to help them gather, organize, analyze, and present their information. This plan will help you use scientific inquiry and process skills to work like a scientist.

## Step 1—Observe and ask questions.

Which soil works best for planting marigold seeds?

- Use your senses to make observations.
- Record a question you would like to answer.

## Step 2—Make a hypothesis.

My hypothesis: Marigold seeds sprout best in potting soil.

- Choose one possible answer, or hypothesis, to your question.
- Write your hypothesis in a complete sentence.
- Think about what investigation you can do to test your hypothesis.

## Step 3—Plan your test.

I'll put identical seeds in three different kinds of soil.

- Write down the steps you will follow to do your test. Decide how to conduct a fair test by controlling variables.
- Decide what equipment you will need.
- Decide how you will gather and record your data.

## Step 4—Conduct your test.

I'll make sure to record my observations each day. Each flowerpot will get the same amount of water and light.

- Follow the steps you wrote.
- Observe and measure carefully.
- Record everything that happens.
- Organize your data so that you can study it carefully.

## Step 5—Draw conclusions and share results.

Hmm. My hypothesis was not correct. The seeds sprouted equally well in potting soil and sandy soil. They didn't sprout at all in clay soil.

- Analyze the data you gathered.
- Make charts, graphs, or tables to show your data.
- Write a conclusion. Describe the evidence you used to determine whether your test supported your hypothesis.
- Decide whether your hypothesis was correct.

### Investigate Further

I wonder if a combination of soils would work best. Maybe I'll try...

# Using Science Tools

## Using a Hand Lens

**A hand lens magnifies objects, or makes them look larger than they are.**

1. Hold the hand lens about 12 centimeters (5 in.) from your eye.

2. Bring the object toward you until it comes into focus.

## Using a Thermometer

**A thermometer measures the temperature of air and most liquids.**

1. Place the thermometer in the liquid. Don't touch the thermometer any more than you need to. Never stir the liquid with the thermometer. If you are measuring the temperature of the air, make sure that the thermometer is not in line with a direct light source.

2. Move so that your eyes are even with the liquid in the thermometer.

3. If you are measuring a material that is not being heated or cooled, wait about two minutes for the reading to become stable, or stay the same. Find the scale line that meets the top of the liquid in the thermometer, and read the temperature.

4. If the material you are measuring is being heated or cooled, you will not be able to wait before taking your measurements. Measure as quickly as you can.

# Caring for and Using a Microscope

A microscope is another tool that magnifies objects. A microscope can increase the detail you see by increasing the number of times an object is magnified.

### Caring for a Microscope

- Always use two hands when you carry a microscope.
- Never touch any of the lenses of a microscope with your fingers.

### Using a Microscope

1. Raise the eyepiece as far as you can by using the coarse-adjustment knob. Place your slide on the stage.

2. Always start by using the lowest power. The lowest-power lens is usually the shortest. Start with the lens in the lowest position it can go without touching the slide.

3. Look through the eyepiece, and begin adjusting it upward with the coarse-adjustment knob. When the slide is close to being in focus, use the fine-adjustment knob.

4. When you want to use a higher-power lens, first focus the slide under low power. Then, watching carefully to make sure that the lens will not hit the slide, turn the higher-power lens into place. Use only the fine-adjustment knob when looking through the higher-power lens.

You may use a Brock microscope. This is a sturdy microscope that has only one lens.

1. Place the object to be viewed on the stage.

2. Look through the eyepiece, and begin raising the tube until the object comes into focus.

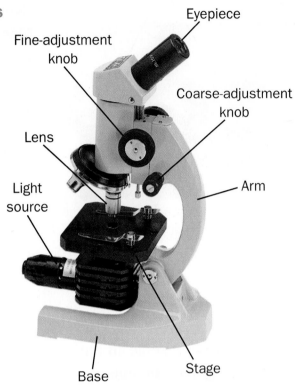

**A Light Microscope**

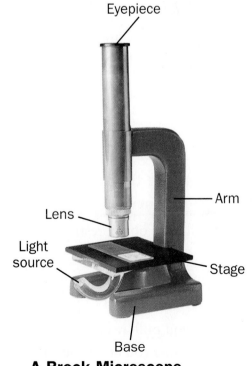

**A Brock Microscope**

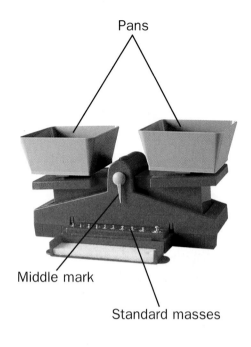

Pans

Middle mark

Standard masses

## Using a Balance

**Use a balance to measure an object's mass. Mass is the amount of matter an object has.**

1. Look at the pointer on the base to make sure the empty pans are balanced.

2. Place the object you wish to measure in the left-hand pan.

3. Add the standard masses to the other pan. As you add masses, you should see the pointer move. When the pointer is at the middle mark, the pans are balanced.

4. Add the numbers on the masses you used. The total is the mass in grams of the object you measured.

## Using a Spring Scale

**Use a spring scale to measure forces such as the pull of gravity on objects. You measure weight and other forces in units called newtons (N).**

**Measuring the Weight of an Object**

1. Hook the spring scale to the object.

2. Lift the scale and object with a smooth motion. Do not jerk them upward.

3. Wait until any motion of the spring comes to a stop. Then read the number of newtons from the scale.

**Measuring the Force to Move an Object**

1. With the object resting on a table, hook the spring scale to it.

2. Pull the object smoothly across the table. Do not jerk the object.

3. As you pull, read the number of newtons you are using to pull the object.

## Measuring Liquids

Use a beaker, a measuring cup, or a graduate
to measure liquids accurately.

1. Pour the liquid you want to measure into a measuring container. Put your measuring container on a flat surface, with the measuring scale facing you.

2. Look at the liquid through the container. Move so that your eyes are even with the surface of the liquid in the container.

3. To read the volume of the liquid, find the scale line that is even with the surface of the liquid.

4. If the surface of the liquid is not exactly even with a line, estimate the volume of the liquid. Decide which line the liquid is closer to, and use that number.

**Beaker**          **Graduate**

## Using a Ruler or Meterstick

Use a ruler or meterstick to measure distances and
to find lengths of objects.

1. Place the zero mark or end of the ruler or meterstick next to one end of the distance or object you want to measure.

2. On the ruler or meterstick, find the place next to the other end of the distance or object.

3. Look at the scale on the ruler or meterstick. This will show the distance you want or the length of the object.

## Using a Timing Device

Use a timing device such as a stopwatch to
measure time.

1. Reset the stopwatch to zero.

2. When you are ready to begin timing, press *Start*.

3. As soon as you are ready to stop timing, press *Stop*.

4. The numbers on the dial or display show how many minutes, seconds, and parts of seconds have passed.

# Using a Computer

**A computer can help you communicate with others and can help you get information. It is a tool you can use to write reports, make graphs and charts, and do research.**

## Writing Reports

To write a report with a computer, use a word processing software program. After you are in the program, type your report. By using certain keys and the mouse, you can control how the words look, move words, delete or add words and copy them, check your spelling, and print your report.

Save your work to the desktop or hard disk of the computer, or to a floppy disk. You can go back to your saved work later if you want to revise it.

There are many reasons for revising your work. You may find new information to add or mistakes you want to correct. You may want to change the way you report your information because of who will read it. Computers make revising easy. You delete what you don't want, add the new parts, and then save. You can save different versions of your work if you want to.

For a science lab report, it is important to show the same kinds of information each time. With a computer,
you can make a general format for a lab report, save the format, and then use it again and again.

## Making Graphs and Charts

You can make a graph or chart with most word processing software programs. You can also use special software programs such as Data ToolKit or Graph Links. With Graph Links you can make pictographs and circle, bar, line, and double-line graphs.

First, decide what kind of graph or chart will best communicate your data. Sometimes it's easiest to do this by sketching your ideas on paper. Then you can decide what format and categories you need for your graph or chart. Choose that format

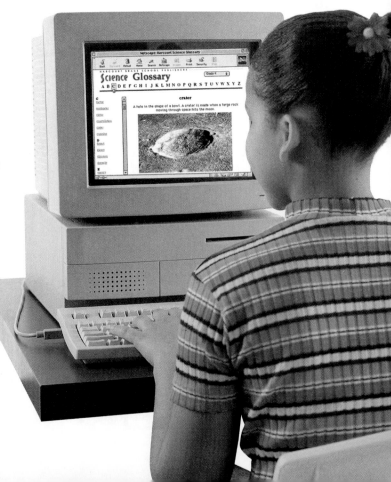

for the program. Then type your information. Most software programs include a tutor that gives you step-by-step directions for making a graph or chart.

### Doing Research

Computers can help you find current information from all over the world through the Internet. The Internet connects thousands of computer sites that have been set up by schools, libraries, museums, and many other organizations.

Get permission from an adult before you log on to the Internet. Find out the rules for Internet use at school or at home. Then log on and go to a search engine, which will help you find what you need. Type in keywords, words that tell the subject of your search. If you get too much information that isn't exactly about the topic, make your keywords more specific. When you find the information you need, save it or print it.

Harcourt Science tells you about many Internet sites related to what you are studying. To find out about these sites, called Web sites, look for Technology Links in the lessons in this book. If you need to contact other people to help in your research, you can use e-mail. Log into your e-mail program, type the address of the person you want to reach, type your message, and send it. Be

sure to have adult permission before sending or receiving e-mail.

Another way to use a computer for research is to access CD-ROMs. These are discs that look like music CDs. CD-ROMs can hold huge amounts of data, including words, still pictures, audio, and video. Encyclopedias, dictionaries, almanacs, and other sources of information are available on CD-ROMs. These computer discs are valuable resources for your research.

# Glossary

This Glossary contains important science words and their definitions. Each word is respelled as it would be in a dictionary.  When you see the ′ mark after a syllable, pronounce that syllable with more force than the other syllables. The page number at the end of the definition tells where to find the word in your book. The boldfaced letters in the examples in the Pronunciation Key that follows show how these letters are pronounced in the respellings after each glossary word.

## PRONUNCIATION KEY

| | | | | | | | |
|---|---|---|---|---|---|---|---|
| a | **a**dd, m**a**p | m | **m**ove, see**m** | u | **u**p, d**o**ne |
| ā | **a**ce, r**a**te | n | **n**ice, ti**n** | û(r) | b**ur**n, t**er**m |
| â(r) | c**a**re, **ai**r | ng | ri**ng**, so**ng** | yo͞o | f**u**se, f**ew** |
| ä | p**a**lm, f**a**ther | o | **o**dd, h**o**t | v | **v**ain, e**v**e |
| b | **b**at, ru**b** | ō | **o**pen, s**o** | w | **w**in, a**w**ay |
| ch | **ch**eck, cat**ch** | ô | **o**rder, j**aw** | y | **y**et, **y**earn |
| d | **d**og, ro**d** | oi | **oi**l, b**oy** | z | **z**est, mu**s**e |
| e | **e**nd, p**e**t | ou | p**ou**t, n**ow** | zh | vi**s**ion, plea**s**ure |
| ē | **e**qual, tr**ee** | o͝o | t**oo**k, f**u**ll | ə | the schwa, an |
| f | **f**it, hal**f** | o͞o | p**oo**l, f**oo**d | | unstressed vowel |
| g | **g**o, lo**g** | p | **p**it, sto**p** | | representing the sound |
| h | **h**ope, **h**ate | r | **r**un, poo**r** | | spelled |
| i | **i**t, g**i**ve | s | **s**ee, pa**ss** | | *a* in **a**bove |
| ī | **i**ce, wr**i**te | sh | **s**ure, ru**sh** | | *e* in sick**e**n |
| j | **j**oy, le**dge** | t | **t**alk, si**t** | | *i* in poss**i**ble |
| k | **c**ool, ta**k**e | th | **th**in, bo**th** | | *o* in mel**o**n |
| l | **l**ook, ru**l**e | t͟h | **th**is, ba**th**e | | *u* in circ**u**s |

Other symbols:
- • separates words into syllables
- ′ indicates heavier stress on a syllable
- ′ indicates light stress on a syllable

**acceleration** [ak•sel′ər•ā′shən] A change in the speed or direction of an object's motion **(F14)**

**adaptation** [ad′əp•tā′shən] A body part or behavior that helps an animal meet its needs in its environment **(A40)**

**air mass** [âr′mas′] A huge body of air which all has similar temperature and moisture **(D13)**

**air pressure** [âr′presh′ər] Particles of air pressing down on the Earth's surface **(D7)**

**amphibian** [am•fib′ē•ən] An animal that has moist skin and no scales **(A12)**

**anemometer** [an′ə•mom′ə•tər] An instrument used to measure wind speed **(D21)**

**artery** [är′tər•ē] A blood vessel that takes blood away from the heart **(A97)**

**arthropod** [är′thrə•pod] An invertebrate with legs that have several joints **(A13)**

**asteroid** [as′tə•roid] A small rocky object that moves around the sun **(D57)**

**atmosphere** [at′məs•fir] The layer of air that surrounds our planet **(D6)**

**axis** [ak′sis] An imaginary line which runs through both poles of a planet **(D58)**

**barometer** [bə•rom′ət•ər] An instrument that measures air pressure **(D20)**

**brain** [brān] The control center of your nervous system **(A102)**

**buoyancy** [boi′ən•sē] The ability of matter to float in a liquid or gas **(E20)**

**camouflage** [kam′ə•fläzh′] An animal's color or pattern that helps it blend in with its surroundings **(A44)**

**capillary** [kap′ə•ler′ē] A tiny blood vessel that allows gases and nutrients to pass from blood to cells **(A96)**

**carbon dioxide** [kär′bən dī•ok′sīd′] A gas breathed out by animals **(A64)**

**cardiac muscle** [kär′dē•ak mus′əl] A type of muscle that works the heart **(A91)**

**cast** [kast] A fossil formed when sediments or minerals fill a mold; it takes on the same outside shape as the living thing that shaped the mold **(C64)**

**cell** [sel] The basic building block of life **(A90)**

**charge** [chärj] A measure of the extra positive or negative particles that an object has **(E90)**

**circuit** [sûr′kit] A path that is made for an electric current **(E96)**

**classification** [klas′ə•fə•kā′shən] The grouping of things by using a set of rules **(A6)**

**climate** [klī′mit] The average temperature and rainfall of an area over many years **(A33, B26)**

**comet** [kom′it] A small mass of dust and ice that orbits the sun in a long, oval-shaped path **(D57)**

**community** [kə•myoo′nə•tē] All the populations that live in the same area **(B14)**

**compression** [kəm•presh′ən] The part of a sound wave in which air is pushed together **(E63)**

**condensation** [kon′dən•sā′shən] The process by which water vapor changes from a gas to liquid **(D34)**

**conduction** [kən•duk′shən] The transfer of thermal energy caused by particles of matter bumping into each other **(E41)**

**conductor** [kən•duk′tər] A material that electric current can pass through easily **(E97)**

**conservation** [kon′sər•vā′shən] The careful management and wise use of natural resources **(B86)**

**constellation** [kon′stə•lā′shən] The pattern formed by a group of stars in the sky **(D78)**

**consumer** [kən•soo′mər] A living thing that eats other living things for energy **(B21)**

**contour plowing** [kon′toor plou′ing] Plowing around a hill to reduce erosion **(B56)**

**convection** [kən•vek′shən] The transfer of thermal energy by particles of a liquid or gas moving from one place to another **(E42)**

**core** [kôr] The dense center of Earth; a ball made mostly of two metals, iron and nickel **(C6)**

**crater** [krā′tər] A large basin formed at the top of a volcano when it falls in on itself **(C20)**

**crust** [krust] Earth's outer layer; includes the rock of the ocean floor and large areas of land **(C6)**

**decomposer** [dē′kəm•pōz′ər] A living thing that feeds on the wastes of plants and animals or on their remains after they die **(B21)**

**deep ocean current** [dēp′ ō′shən kûr′ənt] An ocean current formed when cold water flows underneath warm water **(D44)**

**density** [den′sə•tē] The property of matter that compares the amount of matter to the space it takes up **(E14)**

**dissolve** [di•zolv′] To form a solution with another material **(E19)**

**diversity** [di•vûr′sə•tē] Variety **(B27)**

**earthquake** [ûrth′ kwāk′] A vibration, or shaking, of Earth's crust **(C12)**

**echo** [ek′ō] A sound reflection **(E76)**

**ecosystem** [ek′ō•sis′təm] Groups of living things and the environment they live in **(B12)**

**effort force** [ef′ərt fôrs′] The force put on one part of the bar when you push or pull on a lever **(F38)**

**electric cell** [i·lek′trik sel′] A battery that supplies energy to move charges through a circuit **(E96)**

**electric current** [i·lek′trik kûr′ənt] A flow of electric charges **(E96)**

**electric field** [i·lek′trik fēld′] The space around an object in which electric forces occur **(E92)**

**electromagnet** [i·lek′trō·mag′nit] An arrangement of wire wrapped around a core, producing a temporary magnet **(E109)**

**energy** [en′ər·jē] The ability to cause a change **(E34)**

**environment** [in·vī′rən·mənt] Everything that surrounds and affects an animal, including living and nonliving things **(A32)**

**epicenter** [ep′i·sent′ər] The point on the surface of Earth that is right above the focus of an earthquake **(C13)**

**erosion** [i·rō′zhən] The process by which wind and moving water carry away bits of rock **(B43)**

**esophagus** [i·sof′ə·gəs] The tube that connects your mouth with your stomach **(A104)**

**evaporation** [ē·vap′ə·rā′shən] The process in which a liquid changes to a gas **(D34)**

**fault** [fôlt] A break in Earth's crust along which rocks move **(C12)**

**fertile** [fûr′təl] Soil that has the nutrients to grow many plants **(B50)**

**fibrous roots** [fī′brəs rōōts′] Long roots that grow near the surface **(A71)**

**focus** [fō′kəs] The point underground where the movement of an earthquake first took place **(C13)**

**force** [fôrs] A push or pull **(F12)**

**fossil** [fos′əl] A preserved clue to life on Earth long ago **(C62)**

**frame of reference** [frām′ uv ref′ər·əns] The things around you that you can sense and use to describe motion **(F7)**

**friction** [frik′shən] A force that keeps objects that are touching each other from sliding past each other easily **(F26)**

**front** [frunt] The border where two air masses meet **(D14)**

**fuel** [fyōō′əl] A material that can burn **(E48)**

**fulcrum** [fŏŏl′krəm] The fixed point, or point that doesn't move, on a lever **(F38)**

**fungi** [fun′ji′] Living things such as mushrooms that look like plants, but can not make their own food **(A7)**

**gas** [gas] The state of matter that has no definite shape and takes up no definite amount of space **(E8)**

**gas giants** [gas′ jī′ənts] The planets Jupiter, Saturn, Uranus, and Neptune, which are large spheres made up mostly of gases **(D64)**

**genus** [jē′nəs] The second-smallest name grouping used in classification **(A8)**

**germinate** [jûr′mə•nāt′] To sprout; said of a seed **(A76)**

**gravity** [grav′ə•tē] A force that pulls all objects toward each other **(F22)**

**greenhouse effect** [grēn′hous′ i•fekt′] The warming of Earth caused by the atmosphere trapping thermal energy from the sun **(D12)**

**habitat** [hab′ə•tat′] An environment that meets the needs of an organism **(B20)**

**hardness** [härd′nis] A mineral's ability to resist being scratched **(C35)**

**heart** [härt] The muscle that pumps blood through your blood vessels to all parts of your body **(A97)**

**heat** [hēt] The transfer of thermal energy from one piece of matter to another **(E40)**

**hibernation** [hī′bər•nā′shən] A period when an animal goes into a long, deep "sleep" **(A51)**

**humidity** [hyōō•mid′ə•tē] The amount of water vapor in the air **(D21)**

**humus** [hyōō′məs] The rotting plant and animal materials in topsoil **(B44)**

**igneous rock** [ig′nē•əs rok′] A rock that forms when completely melted rock hardens **(C40)**

**inclined plane** [in•klīnd′ plān′] A flat surface with one end higher than the other **(F52)**

**infrared radiation** [in′frə•red′ rā′dē•ā′shən] The bundles of light energy that transfer heat **(E44)**

**inner planets** [in′ər plan′its] The planets closest to the sun; Mercury, Venus, Earth, and Mars **(D62)**

**instinct** [in′stingkt] A behavior that an animal begins life with **(A48)**

**insulator** [in′sə•lāt′ər] A material that current cannot pass through easily **(E97)**

**invertebrate** [in•vûr′tə•brit] An animal without a backbone **(A13)**

**kingdom** [king′dəm] The largest group into which living things can be classified **(A7)**

**large intestine** [lärj in•tes′tən] The last part of the digestive system where water is removed from food **(A105)**

**lava** [lä′və] A melted rock that reaches Earth's surface **(C18)**

**lever** [lev′ər] A simple machine made up of a bar that turns around a fixed point **(F38)**

**liquid** [lik′wid] The state of matter that takes the shape of its container and takes up a definite amount of space **(E7)**

**loudness** [loud′nes] A measure of the amount of sound energy reaching your ear **(E68)**

**lungs** [lungz] The main organs of the respiratory system **(A96)**

**luster** [lus′tər] A way that the surface of a mineral reflects light **(C35)**

**magma** [mag′mə] Melted rock inside Earth **(C18)**

**magma chamber** [mag′mə chām′bər] An underground pool below a volcano that holds magma **(C19)**

**magnet** [mag′nit] An object that attracts certain materials, such as iron or steel **(E102)**

**magnetic field** [mag•net′ik fēld′] The space all around a magnet where the force of the magnet can act **(E102)**

**magnetic pole** [mag•net′ik pōl′] The end of a magnet **(E102)**

**mammal** [mam′əl] An animal that has hair and produces milk for its young **(A12)**

**mantle** [man′təl] The thickest layer of Earth; found just below the crust **(C6)**

**mass** [mas] The amount of matter something contains **(E6)**

**matter** [mat′ər] Everything in the universe that has mass and takes up space **(E6)**

**metamorphic rock** [met′ə•môr′fik rok′] A rock changed by heat or pressure, but not completely melted **(C44)**

**metamorphosis** [met′ə•môr′fə•sis] The process of change; for example, from an egg to an adult butterfly **(A36)**

**migration** [mī•grā′shən] The movement of a group of one type of animal from one region to another and back again **(A49)**

**mimicry** [mim′ik•rē] An adaptation in which an animal looks very much like another animal or an object **(A44)**

**mineral** [min′ər•əl] A natural, solid material with particles arranged in a repeating pattern **(C34)**

**mold** [mōld] A fossil imprint made by the outside of a dead plant or animal **(C64)**

**mollusk** [mol′əsk] An invertebrate that may or may not have a hard outer shell **(A13)**

**moneran** [mō•ner′ən] The kingdom of classification for organisms that have only one cell **(A7)**

**motion** [mō′shən] A change of position **(F6)**

**nerve** [nûrv] A group of neurons that carries signals from the brain to the body and from the body to the brain **(A102)**

**neuron** [noor′on′] A nerve cell **(A102)**

**newton** [nōō′tən] The metric, or Système International (SI), unit of force **(F17)**

**niche** [nich] The role or part played by an organism in its habitat **(B21)**

**nonvascular plants** [non•vas′kyə•lər plants′] The plants that do not have tubes **(A20)**

**nutrient** [nōō′trē•ənt] Substances, such as minerals, which all living things need in order to grow **(A64)**

**orbit** [ôr′bit] The path that an object such as a planet makes as it revolves around a second object **(D58)**

**organ** [ôr′gən] A group of tissues of different kinds working together to perform a task **(A90)**

**outer planets** [ou′tər plan′its] The planets farthest from the sun; Jupiter, Saturn, Uranus, Neptune, and Pluto **(D64)**

**oxygen** [ok′si•jən] One of the many gases in air **(A33)**

**parallel circuit** [par′ə•lel sûr′kit] A circuit that has more than one path along which current can travel **(E98)**

**photosynthesis** [fōt′ō•sin′thə•sis] The process by which a plant makes its own food **(A65)**

**pitch** [pich] A measure of how high or low a sound is **(E69)**

**planet** [plan′it] A large object that moves around a star **(D57)**

**plate** [plāt] Continent-sized slab of Earth's crust and upper mantle **(C8)**

**population** [pop′yə•lā′shən] A group of the same species living in the same place at the same time **(B13)**

**position** [pə•zish′ən] A certain place **(F6)**

**precipitation** [pri•sip′ə•tā′shən] Water that falls to Earth as rain, snow, sleet, or hail **(D35)**

**preservation** [prez′ər•vā′shən] The protection of an area **(B88)**

**producer** [prə•dōōs′ər] Living things such as plants that produce their own food **(B21)**

**protist** [prō′tist] The kingdom of classification for organisms that have only one cell and also have a nucleus or cell control center **(A7)**

**pulley** [pŏŏl′ē] A simple machine made up of a rope or chain and a wheel around which the rope fits **(F46)**

**radiation** [rā′dē•ā′shən] The bundles of energy that move through matter and through empty space **(E44)**

**reclamation** [rek′lə•mā′shən] The repairing of some of the damage done to an ecosystem **(B81)**

**relative motion** [rel′ə•tiv mō′shən] A motion that is described based on a frame of reference **(F7)**

**reptile** [rep′təl] An animal that has dry, scaly skin **(A12)**

**resistance force** [ri•zis′təns fôrs′] The force put out by the other end of the bar on a lever; the force that does work for you **(F38)**

**resistor** [ri•zis′tər] A material that resists the flow of current but doesn't stop it **(E97)**

**rock** [rok] A material made up of one or more minerals **(C40)**

**rock cycle** [rok′ sī′kəl] The slow, never-ending process of rock changes **(C50)**

**S**

**salinity** [sə•lin′ə•tē] The amount of salt in water **(B28)**

**screw** [skro͞o] An inclined plane wrapped around a pole **(F54)**

**sedimentary rock** [sed′ə•men′tər•ē rok′] A rock formed by layers of sediments squeezed and stuck together over a long time **(C42)**

**seismograph** [sīz′mə•graf′] An instrument that records earthquake waves **(C14)**

**series circuit** [sir′ēz sûr′kit] A circuit that has only one path for current **(E98)**

**shelter** [shel′tər] A place where an animal is protected from other animals or from the weather **(A35)**

**simple machine** [sim′pəl mə•shēn′] One of the basic machines that make up other machines **(F38)**

**small intestine** [smôl′ in•tes′tən] A long tube of muscle where most food is digested **(A104)**

**smooth muscle** [smo͞oth′ mus′əl] A type of muscle found in the walls of some organs such as the stomach, intestines, blood vessels, and bladder **(A92)**

**soil conservation** [soil′ kon′sər•vā′shən] The saving of soil **(B56)**

**solar energy** [sō′lər en′ər•jē] The energy given off by the sun **(E49)**

**solar system** [sō′lər sis′təm] A group of objects in space that move around a central star **(D56)**

**solid** [sol′id] The state of matter that has a definite shape and takes up a definite amount of space **(E6)**

**solubility** [sol′yə•bil′ə•tē] A measure of the amount of a material that will dissolve in another material **(E19)**

**solution** [sə•lo͞o′shən] A mixture in which the particles of different kinds of matter are mixed evenly with each other and particles do not settle out **(E18)**

**sonic boom** [son′ik bo͞om′] A shock wave of compressed sound waves produced by an object moving faster than sound **(E78)**

**sound** [sound] A series of vibrations that you can hear **(E62)**

**sound wave** [sound′ wāv′] A moving pattern of high and low pressure that you can hear **(E63)**

**space probe** [spās′ prōb′] A space vehicle that carries cameras, instruments, and other research tools **(D74)**

**species** [spē′shēz] The smallest name grouping used in classification **(A8)**

**speed** [spēd] A measure of an object's change in position during a unit of time; for example, 10 meters per second **(F8)**

**speed of sound** [spēd′ uv sound′] The speed at which a sound wave travels through a given material **(E74)**

**spinal cord** [spi′nəl kôrd′] The tube of nerves that runs through your spine, or backbone **(A102)**

**spore** [spôr] A tiny cell that ferns and fungi use to reproduce **(A77)**

**stability** [stə•bil′ə•tē] The condition that exists when the changes in a system over time cancel each other out **(B8)**

**star** [stär] A huge, burning sphere of gases; for example, the sun **(D56)**

**static electricity** [stat′ik ē′lek•tris′i•tē] An electric charge that stays on an object **(E90)**

**stomach** [stum′ək] A bag made up of smooth muscles that mixes food with digestive juices **(A104)**

**storm surge** [stôrm′ sûrj′] A very large wave caused by high winds over a large area of ocean **(D41)**

**stratosphere** [strat′ə•sfir′] The layer of atmosphere that contains ozone and is located above the troposphere **(D8)**

**streak** [strēk] The color of the powder left behind when you rub a mineral against a white tile called a streak plate **(C35)**

**striated muscle** [strī′āt•ed mus′əl] A muscle with light and dark stripes; a muscle you can control by thinking **(A92)**

**strip cropping** [strip′ krop′ing] The practice of planting one or more crops between rows of other crops to control erosion **(B56)**

**succession** [sək•sesh′ən] The process that gradually changes an existing ecosystem into another ecosystem **(B70)**

**surface current** [sûr′fis kûr′ənt] An ocean current formed when steady winds blow over the surface of the ocean **(D44)**

**symmetry** [sim′ə•trē] The condition in which each feature on one half of an object has a matching feature on the other half **(A70)**

**system** [sis′təm] A group of parts that work together as a unit **(B6)**

**taproot** [tap′rōōt′] A plant's single main root that goes deep into the soil **(A71)**

**telescope** [tel′ə•skōp′] A device people use to observe distant objects with their eyes **(D70)**

**temperature** [tem′pər•ə•chər] A measure of the average energy of motion of the particles in matter **(E35)**

**terracing** [ter′əs•ing] A farming method used on steep hillsides to control erosion **(B56)**

**thermal energy** [thûr′məl en′ər•jē] The energy of the motion of particles in matter **(E34)**

Multimedia Science Glossary: **www.harcourtschool.com/scienceglossary**

**tide** [tīd] The daily changes in the local water level of the ocean **(D42)**

**tissue** [tish′o͞o] A group of cells of the same type **(A90)**

**trace fossil** [trās′ fos′əl] A fossil that shows changes that long-dead animals made in their surroundings **(C63)**

**transpiration** [tran′spə•rā′shən] The giving off of water vapor by plants **(A70)**

**troposphere** [trō′pə•sfir′] The layer of atmosphere closest to Earth **(D8)**

**tuber** [to͞o′bər] A swollen underground stem **(A78)**

**vascular plant** [vas′kyə•lər plant′] A plant that has tubes **(A18)**

**vein** [vān] A large blood vessel that returns blood to the heart **(A97)**

**vent** [vent] The rocky tube in a volcano through which magma rises toward the surface **(C18)**

**vertebrate** [vûr′tə•brit] An animal with a backbone **(A12)**

**volcano** [vol•kā′nō] A mountain that forms when red-hot melted rock flows through a crack onto Earth's surface **(C18)**

**volume** [vol′yəm] The amount of space that matter takes up **(E13)**

**water cycle** [wôt′ər sī′kəl] The constant recycling of water on Earth **(D34)**

**wave** [wāv] The up-and-down movement of water **(D40)**

**weathering** [weth′ər•ing] The process by which rocks are broken down into smaller pieces **(B43, C42)**

**wedge** [wej] A machine made up of two inclined planes placed back-to-back **(F56)**

**weight** [wāt] A measure of the force of gravity upon an object **(F23)**

**wheel and axle** [hwēl′ and ak′səl] A simple machine made up of a large wheel attached to a smaller wheel or rod **(F48)**

**work** [wûrk] That which is done on an object when a force moves the object through a distance **(F42)**

**A**

Abdominal muscles, R32
Academy of Natural
 Sciences, B92
Acceleration, F14
Acoustic engineer, E81
Active noise control (ANC),
 E80-81
Activity pyramid, planning
 weekly, R16
Adams, John C., D83
Adaptation(s), B13
 animal, A40-45
 defined, A40
Adhesives, medical, A107
African elephants, G68-69
Agricola, Georgius, C74
Air
 properties of, D6-7
 and sun, D12
 and weather, D12–17
Air conditioning, E52–53
Air masses, D13
 meeting of, D14–15
 movement of, D16–17
Air pressure, D7, D16
 and sonic booms, E78
Air sacs, A96
Alarm bell, workings of,
 E110–111
Aldabra tortoise, A29
Aldrin, Edwin "Buzz," D73
Algae, B71
Alligators, A44
Amber, E114
American Geophysical
 Union, B62
Ammonia, B50, B61, E53
Ammonoids, C59
Amphibians, A12
Andes, C18
Andesite, C56
Anemometer, D21

Animal(s)
 African, A34
 behavior of, A48–52
 body coverings of, A42–43
 color and shape of, A44
 described, A12
 needs of, A32–37
 speed of, A28
 what they eat, A34
 young of, A36–37
Animal behaviorist, A56
Animal kingdom, A7
Animatronic dinosaur, C71
Anvil, E64
Apache families, B60–61
*Apollo 11*, D73
Archerfish, B14
Arches National Park, UT, C38
Archimedes screw, F50–51, F54
Arecibo, Puerto Rico, D71
Aristotle, A24
Armstrong, Neil, D73
Arteries, A97, A98, R36
Arthropods, A18
Asteroids, D57
Astronaut, F30
Astronomers, D71, D84
Astronomical units (AU), D60
Atom(s), F24–25
Atomic forces, F25
Attini ants, B39
Attraction, E92
Auger, F55
Axis, D58

**B**

Babbage, Charles, B35
Backbone, A91
Bacteria, A6, B71
 and food safety, R14
Balance, ecosystem, B70, B78
Balance, measuring, R6
Balanced forces, F16
Balanced lever, F38, F39, F40
Bark, A19
Barometer, D20
Basalt, C41, C48
Bat(s), North American, A51
 hunting technique of, E58
Bathyscaphe, D46
Battery, invention of, E115
Bay of Fundy, Canada, D42
Beaker, E13
Beaks, bird, A40
Bedrock, B44
Beggar's-lice seeds, D79
Behaviors, learned, A52–53
Berson, Solomon, A108
Beta carotene, A81
BetaSweet carrot, A81
Biceps, A91, A92, R32
Bicycle mechanic, F29
Big Dipper, D79
Bioremediation specialist, B91
Bird(s)
 adaptations of, A40–41
 sounds, E68
Bison, American, A43
Block and tackle, F59
Blood, paths of, A98
Blood vessels, R36
Blow torch, E40
Boat oar, as lever, F36
Boiling, E8, E34
Bone meal, B60–61

**L**

 **M**

Motor nerves, R40
Motors, electric, E112
Mountains, formation of, C8
Mount Fuji, Japan, C19, C21
Mount Mazama, C20
Mount Pelée, Martinique, C19
Mount St. Helens, WA, C22, C23
Mount Vesuvius, Italy, C19, C21
Mouth, R34, R38
Movable pulley, F46–47
MRI machines, E115
Mucus, R29
Mulberry leaves, A70
Muscles, human, A14, A15, A90
Muscular system, A90–91,
    R32–33
    caring for, R33
Music scale, E65
Mute, E70

NASA, C54, D26, F30
Nasal cavity, R29
National Aeronautic and Space
    Administration. *See* NASA
National Herbarium,
    Smithsonian, A82
National Oceanic and
    Atmospheric Adminstration
    (NOAA), D26
National parks (chart), B88
National parks and forests, B88
National Park Service, B34
National Physics
    Laboratory, E82
Native Americans, planting
    methods, B60
*Nature's Great Balancing Act*
    (Norsgaard), B17
Nautiloid, C60
Neap tides, D43
Negative charge, E90, E91–93,
    F24, F25
Neptune, D65, D83
Nerves, A102–103

Nervous system, A102–103,
    R40–41
    caring for, R40
Neurons, A102
Neutral charges, E91
Neutrons, F24
Newton (N), F17
Niches, B21–22
Nitrogen, B58, B60, B61
Nobel Prize for Medicine, A108
Noise, E80–81
Noise reduction, E80–81
Nonvascular plants, A20–21
North America, air masses of,
    D13
North Pole, Earth's, D79
North-seeking pole, E102
Nose, R38
    caring for, R29
Nostrils, R29
Nuclear fission, F25
Nucleus, atomic, F24
Nucleus, cell, A7
Nutrients, A64, A104
    controlling loss of, in soil,
    B58–59
    soil (chart), B58

Observatory, D71
Obsidian, C41
Ocean(s), comparative sizes of,
    D31
Ocean Drilling Program, C76
Ocean movements, D40–45
Ocean water, D34–37
    composition of, D36–37
Ochoa, Ellen, F30
Oersted, Hans, E115
Oil spill, B76
Olfactory bulb, R29
Olfactory tract, R29
Olympus Mons, D63
Onnes, Heike, E115
Open system, B7

Optical telescope, D70
Optic nerve, R28
Orbits, D58
Order, A8
Organic chemist, C53
Organs, human, A90
Orion, D78, D80
Osprey, A41
    nest of, B82
Ostriches, A28, A41
Outer core, Earth's, C7
Outer ear, E64, R28
Outer planets, D64–66
Outputs, systems, B7
Owen, Richard, C74
Oxygen, A33, A64, A65, A98
Ozone layer, E53

Palissy, Bernard, C74
Pan balance, E12
Parallel circuit, E98
Parent rock, B44
Paricutín volcano, C2
Particles, E90
    of air, D7, D20
    atomic, F24
    in gases, E8
    of heated matter, E41
    in liquids, E7
    in matter, E34
    radioactive, A108
    of solids, E6
Pasta maker, as wheel
    and axle, F44
Patrick, Ruth, B92
Pelvis, A91, R30, R31
Penguin, A36
Permanent magnet, E110
Petrified Forests, AZ, C64–65
Petrified fossils, C64
Phobos, D67
Phoenicians, F58
Phosphorus, B61
Photosynthesis, A65, D7, E49

PHOTO CREDITS:

**Page Placement Key: (t)-top (c)-center (b)-bottom (l)-left (r)-right (fg)-foreground (bg)-background**

**Cover**

Tim Flach/Tony Stone Images; (bg) Richard Price/FPG International

**Contents**

Page: iv Tom McHugh/Steinhart Aquarium/Photo Researchers; iv (bg) Mark Lewis/Liaison International; v Steve Kaufman/DRK Photo; v (bg) H.Richard Johnston; vi Francis Gohier/Photo Researchers; vi (bg) Kunio Owaki/The Stock Market; vii Sp. Bob Gatkany/Dorling Kindersley; vii (bg)Telegraph Colour Library/FPG International; viii Steve Taylor/Tony Stone Images; viii (bg) Michael Abbey/Photo Researchers; ix Steve Berman/Liaison International.

**Unit A**

A2-A3 Gregory Ochocki/Photo Researchers.; A3 (t) Dave Watts/Tom Stack & Associates; A3 (b) Frances Fawcett/Cornell University/American Indian Program; A4 Christian Grzimek/Okapia/Photo Researchers; A6 (l) MESZA/Bruce Coleman, Inc.; A6 (r) Andrew Syred/Science Photo Library/Photo Researchers; A6 (c) Robert Brons/ BPS/ Tony Stone Images; A6-A7Bill Lea/Dembinsky Photo Associates; A7 (t) Bill Lea/ Dembinsky Photo Associates; A7 (c) S. Nielsen/Bruce Coleman; A7 (b) Andrew Syred/ Science Photo Library/ Photo Researchers; A7 (tc) Dr. E.R. Degginger/Color-Pic; A7 (bc) Robert Brons/ BPS/ Tony Stone Images; A9 Daniel Cox/Tony Stone Images; A10 Arthur C. Smith, III/Grant Heilman Photography; A12-A13(b) Runk/Schoenberger/Grant Heilman Photography; A12 (tl) Ana Laura Gonzalez/Animals Animals; A12 ((bl)) Tom Brakefield/The Stock Market; A13 (tl) Leonard Lee Rue, III/Bruce Coleman, Inc.; A13 (tc) Hans Pfletschinger/Peter Arnold, Inc.; A13 (tr) Mark Moffett/Minden Pictures; A13 (br) Larry Lipsky/DRK; A14 (t) James Balog/Tony Stone Images; A14 (b) Stephen Dalton/Photo Researchers; A16 Darrell Gulin/Tony Stone Images; A18 (l) Dr. E.R. Degginger,FPSA/Color-Pic; A19 Phil A. Dotsen/Photo Researchers; A20 (t) Heather Angel/Biofotos; A20-A21 Runk/ Schoenberger/ Grant Heilman Photography; A22 (tl) Art Resource, NY; A22 (tr) The Granger Collection, New York; A22 (bl) Superstock; A22 (br) E. R. Degginger/Photo Researchers; A23 (i) Leonard Lee Rue III/Photo Researchers; A23 S. J. Krasemann/Peter Arnold, Inc.; A24 (t) Courtesy of Hunt Institute for Botanical Documentation, Carnegie Mellon University, Pittsburg, PA.; A24(b) Grant Heilman Photography; A28-A29Manoj Shah/Tony Stone Images; A29 (t) Rudie Kuiter/Innerspace Visions; A29 (b) Peter Weimann/Animals Animals; A30 Corel; A32 (bg) Rich Reid/Earth Scenes; A32 (li) Larry Minden/Minden Pictures; A32 (ri) Renee Lynn/Photo Researchers; A32 (ci) John Cancalosi/DRK; A33 Daniel J. Cox/ Natural Exposures; A34 (t) T. Kitchen/Natural Selection Stock Photography; A34-A35(b) Bios/Peter Arnold, Inc.; A34 ((bl)) Daniel J. Cox/Natural Selection Stock Photography; A35 (l) Osolinski, S. OSF/Animals Animals; A35 (r) David E. Myers/Tony Stone Images; A35 (li) Stephen Krasemann/Tony Stone Images; A35 (ri) E & P Bauer/Bruce Coleman, Inc.; A36 (l) Ralph Clevenger/ Westlight; A36 (bg) B & C Alexander/Photo Researchers; A36 (i) Dan Suzio Photography; A37 (t) Ben Simmons/The Stock Market; A37 (i) D. Parer & E. Parer-Cook/Auscape; A38 (b) Joe McDonald/Bruce Coleman, Inc.; A40 (l) Robert Lankinen/The Wildlife Collection; A40 (r) Martin Harvey/The Wildlife Collection; A40 (c) Zefa Germany/The Stock Market; A41 (t) Fritz Polking/Dembinsky Photo Associates; A41 (b) Tui De Roy/Minden Pictures; A41 (i) Zig Leszczynski/Animals Animals; A42 (l) Tom and Pat Leeson; A42 (r) Zig Leszczynski/Animals Animals; A42 (c) Fred Bavendam/Peter Arnold, Inc.; A42-A43(b) Stuart Westmorland/Tony Stone Images; A43 (t) Bruce Wilson/Tony Stone Images; A43 (c) Wolfgang Kaehler Photography; A43 ((bl)) Martin Harvey/The Wildlife Collection; A43 (br) Bruce Davidson/Animals Animals; A43 (ti) Bruce Wilson/ Tony Stone Images; A44 (t) Art Wolfe/Tony Stone Images; A44 ((bl)) Stephen Krasemann/Tony Stone Images; A44 (br) Stouffer Prod./Animals Animals; A45Joan Baron/The Stock Market; A46 C. Bradley Simmons/Bruce Coleman, Inc.; A48-A49 Mike Severns/Tony Stone Images; A50 (t) Grant Heilman Photography; A50-A51(b) Daniel J. Cox/Tony Stone Images; A51 (t) Joe McDonald/Animals Animals; A51 ((bl)) Darrell Gulin/Tony Stone Images; A51 (br) J. Foott/Bruce Coleman, Inc.; A52 (t) Tom Brakefield/The Stock Market; A52 (b) Mark Petersen/Tony Stone Images; A53 (t) Darryl Torckler/Tony Stone Images; A54 Katsumi Kasahara/Associated Press; A56 Michael K. Nichols/NGS Image Collection; A59 (l) Ralph Clevenger/ Westlight; A59 (li) Dan Suzio Photography; A60-A61Bertram G. Murray, JR./Animals Animals; A61 (l) Gilbert S. Grant/Photo Researchers; A61 (r) J.A. Kraulis/Masterfile; A62 Christi Carter/Grant Heilman Photography; A64 H. Mark Weidman; A65 Porterfield-Chickering/Photo Researchers; A65 (i) Dr. Jeremy Burgess/Science Photo Library/Photo Researchers; A66 (l) Patti Murray/Earth Scenes; A66 (t) Frans Lanting/Minden Pictures; A67 C.K. Lorenz/Photo Researchers; A68 Dr. E.R. Degginger, FPSA/Color-Pic; A70 (tl) Runk/Schoenberger/Grant Heilman Photography; A70 ((bl)) Runk/Schoenberger/Grant Heilman Photography; A70 (br) Dr. E.R. Degginger/Earth Scenes; A71 (tl) John Kaprielian/Photo Researchers; A71 (tr) Runk/Schoenberger/Grant Heilman Photography; A71 ((bl)) Lefever/Grushow/Grant Heilman Photography; A71 (br) Runk/Schoenberger/Grant Heilman Photography; A72 (r) Bill Lea/Dembinsky Photo Associates; A72 (tl) Kim Taylor/Bruce Coleman, Inc.; A72 ((bl)) Kim Taylor/Bruce Coleman, Inc.; A74 Runk/Schoenberger/Grant Heilman Photography; A76 (br) Gregory K. Scott/Photo Researchers; A77 (t) Ed Reschke/Peter Arnold, Inc.; A77 ((bl)) Gay Bumgarner/Tony Stone Images; A77 (br) Laura Riley/Bruce Coleman, Inc.; A78 (l) Runk/Schoenberger/Grant Heilman Photography; A78 (r) Heather Angel/Biofotos; A79 Heather Angel/Biofotos; A80 James Lyle/Texas A&M University; A81 Chris Rogers/The Stock Market; A82 Hunt Institute for Botanical Documentation/Carnegie Mellon University; A85 (t) Bruce Coleman, Inc.; A85 (b) Breck P. Kent/Animals Animals; A86-A87 Dr. Dennis Kunkel/Phototake; A87 (b) Doug Perrine/Auscape; A88 (tr) Al Lamme/Len/Phototake; A88 (br) Astrid & Hanns-Frieder Michler/Science Photo Library/Photo Researchers; A88 (cr) M. Abbey/Photo Researchers; A90 Biophoto Associates/Science Source/Photo Researchers; A92 (tr) Al Lamme/Len/Phototake; A92 (br) Astrid & Hanns-Frieder Michler/Science Photo Library/Photo Researchers; A92 (cr) M. Abbey/Photo Researchers; A102 Biophoto Associates/Photo Researchers; A106 Courtesy of DERMABOND* Topical Skin Adhesive, trademark of Ethicon, Inc.; A107 Owen Franken/Tony Stone Images; A108 (t) UPI/Corbis; A108 (b) Will and Deni McIntyre/Photo Researchers.

**Unit B**

B2-B3 Kim Heacox/Peter Arnold, Inc.; B3 (l) G. Perkins/Visuals Unlimited; B3 (r) Herbert Schwind/Okapia/Photo Researchers, Inc.; B4 Navaswan/FPG International; B6-B7 D. Logan/H. Armstrong Roberts; B8 (t) Mark E. Gibson/Dembinsky Photo Associates; B8 (b) Larry Lefever/Grant Heilman Photography; B9 (t) Mark E. Gibson/Dembinsky Photo Associates; B10 Dr. E. R. Degginger/Color-Pic; B12 (bg) Doug Cheeseman/Peter Arnold, Inc.; B12 (li) Dr. E. R. Degginger/Color-Pic; B12 (ri) Farrell Grehan/Photo Researchers; B13 (bg) Luiz C. Marigo/Peter Arnold, Inc.; B13 (ri) M. Timothy O'Keefe/Bruce Coleman, Inc.; B13 (li) Bernard Boutrit/Woodfin Camp & Associates; B14 (bg) Jim Steinberg/Photo Researchers; B15 (t) M. Timothy O'Keefe/Bruce Coleman, Inc.; B16 (t) Tom Bean/The Stock Market; B16 (b) Grant Heilman Photography; B18 Ken M. Highfill/Photo Researchers; B20 (bg) Rob Lewine/The Stock Market; B20 (ci) Ken Brate/Photo Researchers; B20 (li) Art Wolfe/Tony Stone Images; B20 (ri) Matt Meadows/Peter Arnold, Inc.; B21 (tl) David M. Phillips/Photo Researchers; B21 (tr) R & J Spurr/Bruce Coleman, Inc.; B21 ((bl)) Dwight R. Kuhn; B21 (br) Institut Pasteur/Phototake; B21 (cl) Dr. E. R. Degginger/Photo Researchers; B21 (cr) Lewis Kemper/Tony Stone Images; B22 (t) David Carriere/Tony Stone Images; B22 (b) E & P Bauer/Bruce Coleman, Inc.; B23 Roy Morsch/The Stock Market; B24 Stuart Westmorland/Tony Stone Images; B27 (t) James Martin/Tony Stone Images; B27 (c) Dr. E. R. Degginger/Color-Pic; B27 (b) Tom McHugh/Photo Researchers; B29 (t) Manfred Kage/Peter Arnold, Inc.; B29 (c) William Townsend, Jr/Photo Researchers; B29 (b) Norbert Wu/The Stock Market; B30 (t) Dr. E. R. Degginger/Color-Pic; B30 (b) Dr. E. R. Degginger/Color-Pic; B31 Tom McHugh/Photo Researchers; B32 Norbert Wu Photography; B33 (t) Oakridge National Laboratory; B33 (b)David Young-Wolff/PhotoEdit; B34 (b) David Muench/Tony Stone Images; B34 (t) Courtesy of Indiana Dunes National Lakeshore/National Park Service; B38-B39 Wolfgang Kaehler Photography; B39 (t) Jeff Foott/Bruce Coleman, Inc.; B39 (b) Mark W. Moffett/Minden Pictures; B40 Rod Planck/Photo Researchers; B42 (t) Lynn M. Stone/Natural History Photography.; B42-B43(b) Holt Studios International (Inga Spence)/Photo Researchers; B46 Dr. E.R. Degginger/Color-Pic; B48 (l) Rob Boudreau/Tony Stone Images.; B48 (r) Gary Braasch/Tony Stone Images; B48 (c) Willard Clay/Tony Stone Images; B50 (b) Grant Heilman/Grant Heilman Photography.; B51 Martha McBride/Unicorn Stock Photos.; B52 Kent & Donna Dannen.; B54-B55 (b) C.C. Lockwood/DRK; B54 (i) Grant Heilman/Grant Heilman Photography.; B55 (t) Simon Fraser/Science Photo Library/Photo Researchers; B55 (c) Jim Richardson/Woodfin Camp & Associates.; B56 (t) Earl Roberge/Photo Researchers; B56 (c) Grant Heilman/Grant Heilman Photography.; B56 (b) Hilarie Kavanagh/Tony Stone Images; B57 (t) Heather Angel/Biofotos; B57 (c) Mark E. Gibson.; B57 (b) G.R.Roberts/G.R. Roberts Photo Library; B58 (t) Norman O. Tomalin/Bruce Coleman, Inc.; B58 (b) H.P. Merten/The Stock Market.; B59 Mark E. Gibson.; B60 (l) Roy Morsch/The Stock Market; B60 (r) Norm Thomas/Photo Researchers; B60 (c) Boltin Picture Library; B61 Hans Reinhard/Bruce Coleman, Inc.; B66 Joel Sartore from Grant Heilman Photography; B66-B67Jeff Foott/Bruce Coleman, Inc.; B67 (b) Roy Toft/Tom Stack & Associates; B68 Gary Braasch/Woodfin Camp & Associates.; B71 Mike Yamashita/Woodfin Camp & Associates; B72 (t) Frank Oberle/Tony Stone Images.; B72 ((bl)) David Woods/The Stock Market.; B72 (bri)Stan Osolinski/Dembinsky Photo Associates.; B72-B73 (b) Jeff Henry/Peter Arnold, Inc.; B73 (t) Milton Rand/Tom Stack & Associates; B73(ti)Joe McDonald/Tom Stack & Associates.; B73 (bi) Stan Osolinski/Dembinsky Photo Associates.; B74 (t) Aneal Vohra/Unicorn Stock Photos; B74 (b) Merrilee Thomas/Tom Stack & Associates.; B74 (bi) Tom Benoit/Tony Stone Images.; B75 Mark E. Gibson/Mark E. Gibson.; B76 Tom Walker/Stock, Boston; B78-B79 (b) Jason Hawkes/Tony Stone Images; B78-B79 (t ) David Harp Photography; B79 (tr) Frans Lanting/Tony Stone Images; B79 (br) Scott Slobodian/Tony Stone Images; B80-B81(b) Paul Chesley/Tony Stone Images.; B80 (t) Mark E. Gibson; B81 (t) J. Lotter/Tom Stack & Associates.; B82 (b) Lori Adamski Peek/Tony Stone Images.; B82 (t ) David R. Frazier; B83 Marc Epstein/Visuals Unlimited; B84 Jim Schwabel/New England Stock Photo; B88 (t) Gary Braasch Photography; B88 (b) Barbara Gerlach/Dembinsky Photo Associates.; B89 (b) SuperStock; B89 Superstock; B90 Earl Roberge/Photo Researchers; B91 Ted Streshinsky/Corbis-Bettmann; B92 AP/Wide World Photos.

**Unit C**

C2-C3 Danny Lehman/Corbis; C3(b) Packwood, R. /Earth Scenes; C4 Jock Montgomery/Bruce Coleman, Inc.; C9 Kenneth Fink/Bruce Coleman, Inc.; C10 Wesley Bocxe/Photo Researchers; C12 Francois Gohier/Photo Researchers; C14 (t) Russell D. Curtis/Photo Researchers; C14 ((bl) ) Tom McHugh/Photo Researchers; C14 (bc) Lee Foster/Bruce Coleman, Inc.; C15 Francois Gohier/Photo Researchers; C16 Dr. E.R. Degginger/Photo Researchers; C20-C21C. C. Lockwood/Bruce Coleman, Inc.; C21 (t) Paolo Koch/Photo Researchers; C22 (t) Photo Researchers; C22 (b) H. Hannejdottir/Photo Researchers; C22 (bg) Ken Sakamoto/Black Star; C23 Krafft-Explorer/Photo Researchers; C24 Bill Ingals/NASA/Sygma; C25 Nancy Simmerman/Tony Stone Images; C26 (b) Thomas Jaggar/NGS Image Collection; C30-C31 Dan Suzio/Photo Researchers; C31 (l) Sam Ogden/Science Photo Library/Photo Researchers; C31 (br) Breck P. Kent/Earth Scenes; C32 The Natural History Museum, London; C34 (r) Dr. E.R. Degginger/Color-Pic; C34 (c) Joy Spurr/Bruce Coleman, Inc.; C34 (tl) Dr. E.R. Degginger, FPSA/Color-Pic.; C34 (bl) Dr. E.R. Degginger/Bruce Coleman, Inc.; C35 (c1) Dr. E.R. Degginger,FPSA/Color-Pic; C35 (c2) Dr. E.R. Degginger/Bruce Coleman, Inc.; C35 (c3) Dr.E.R. Degginger/Bruce Coleman, Inc.; C35 (c5) Dr. E.R. Degginger,FPSA/Color-Pic; C35 (c6) Dr. E.R. Degginger/Color-Pic; C35 (c8) Dr. E.R. Degginger,FPSA/Color-Pic; C35 (c9) Mark A. Schneider/Dembinsky Photo Associates; C35 (c10) Dr.E.R. Degginger/Bruce Coleman, Inc.; C36 (tl) Dr. E.R. Degginger/Color-Pic; C36 ((bl)) Dr. E.R. Degginger/Color-Pic; C36 (br) B. Daemmrich/The Image Works; C36 (cl) Biophoto Associates/Photo Researchers; C36 (cr) Andy Sacks/Tony Stone Images; C38Joe McDonald/Bruce Coleman, Inc.; C40 (t) Dr. E.R. Degginger, FPSA/Color-Pic; C40 (b) Phillip Hayson/Photo

Researchers; C41 (b) Martha McBride/Unicorn Stock Photos; C41 (tl) Dr. E.R. Degginger,FPSA/Color-Pic.; C41 (tr) Breck P. Kent/Earth Scenes; C41 (tcr) Robert Pettit/Dembinsky Photo Associates; C41 (tcl) Dr. E.R. Degginger/Color-Pic; C42 Dr. E.R. Degginger/Color-Pic; C43 (tl) Dr. E.R. Degginger/Color-Pic; C43 (tr) Dr. E.R. Degginger/Color-Pic; C43 (bg) David Bassett/Tony Stone Images; C43 (tcl)Dr. E.R. Degginger/Color-Pic; C43 (tcr)Dr. E.R. Degginger, FPSA/Color-Pic; C44 (t) G. R. Roberts Photo Library; C44 (b) Dr. E.R. Degginger/Color-Pic; C44-C45(tl) Dr. E.R. Degginger/Color-Pic; C45 (t) Dr. E.R. Degginger/Color-Pic; C46 Tom Till/Auscape; C48 Dr. E.R. Degginger/Color-Pic; C49 (t) Dr. E.R. Degginger/Color-Pic; C49 (b) Dr. E.R. Degginger/Color-Pic; C50 (l) Dr. E.R. Degginger,FPSA/Color-Pic.; C50 (r) Dr. E.R. Degginger/Color-Pic; C50-C51(t) Dr. E.R. Degginger/Color-Pic; C52-C53James P. Blair & Victor Boswell/NGS Image Collection; C53Mark Richards/PhotoEdit; C54 (t) Photo Courtesy of Mrs. Alma G. Gipson; C54 (b) Stuart McCall/Tony Stone Images; C58-C59 Gerd Ludwig/Woodfin Camp & Associates; C59 (t) The Westthalian Museum of Natural History.; C59 (b) James Martin/Tony Stone Images; C60 Murray Alcosser/The Image Bank; C62 (t) Beth Davidow/Visuals Unlimited; C62 (b) The Natural History Museum, London; C62-C63(b) William E. Ferguson; C63 (c) The Natural History Museum, London; C63 (tl) Dr. P. Evans/Bruce Coleman Collection; C63 (tr) David J. Sams/Tony Stone Images; C64 (l) William E. Ferguson; C64-C65(t) Bob Burch/Bruce Coleman, Inc.; C64 (l) Jane Burton/Bruce Coleman, Inc.; C64-C65(b) Dr. E. R. Degginger/Bruce Coleman, Inc.; C65 (ti) Dr. E. R. Degginger/Bruce Coleman, Inc.; C66 The Natural History Museum, London; C68 (l) The Natural History Museum, London; C68-C69(bg) Mark J. Thomas/Dembinsky Photo Associates; C68 (li) William E. Ferguson; C69 (t) Runk/Schoenberger/Grant Heilman Photography; C70 (t) Phil Schofield/Tony Stone Images; C70 (c) Phil Degginger/Bruce Coleman, Inc.; C71 (c) Louis Psihoyos/Matrix International, Inc.; C71 (ti) Louie Psihoyos/Matrix International, Inc.; C71 (bi) Louie Psihoyos/Matrix International, Inc.; C72 (t) Ted Clutter/Photo Researchers; C72 (b) J. C. Carton/Bruce Coleman, Inc.; C73 William E. Ferguson; C74 (b) The Granger Collection, New York; C74 (tl) Alinari/Art Resource, NY; C74 (tr) The Granger Collection, New York; C75 (t) Bill Bachman/Photo Researchers; C75 (b) James King-Holmes/Science Photo Library/Photo Researchers; C76 (t) San Francisco State University/Department of Geosciences; C76 (b) AP/Wide World Photos.

**Unit D**

D2-D3 Waren Faidley/International Stock Photography; D3 (t) Bob Abraham/The Stock Market; D3 (b) NRSC Ltd/Science Photo Library/Photo Researchers; D4 Keren Su/Stock Boston; D6 Space Frontiers-TCL/Masterfile; D10 Bruce Watkins/Earth Scenes; D12 Peter Menzel/Stock Boston; D14-D15C. O'Rear/Westlight; D16 (b) Bill Binzen/The Stock Market; D18 J. Taposchaner/FPG International; D20 Sam Ogden/Science Photo Library/Photo Researchers; D21 (t) Breck P. Kent/Earth Scenes; D21 (b) B. Daemmrich/The Image Works; D22 © 1998 Accu Weather; D22 © 1998 Accu Weather D46-D47Ben Margot/AP Photo/Wide World Photos; ; D24 Geophysical Institute, University of Alaska, Fairbanks/NASA; D25 Pat Lanza/Bruce Coleman, Inc.; D26 (t) Clark Atlanta University; D26 (b) NASA/Science Photo Library/Photo Researchers; D30-D31Warren Bolster/Tony Stone Images; D31 (t) Warren Morgan/Corbis; D31 (b) Van Van Sant, Geosphere Project/Planetary Visions/Science Photo Library/Photo; D32 Philip A. Savoie/Bruce Coleman, Inc.; D36 (tr) A. Ramey/Stock Boston; D36 ((bl)) Richard Gaul/FPG International; D38 John Lel/Stock Boston; D40 Dr. E.R. Degginger/Photo Researchers; D41 (t) Fredrik Bodin/Stock Boston; D41 (b) Peter Miller/Photo Researchers; D42 (t ) Francois Gohier/Photo Researchers; D42 (c ) Francois Gohier/Photo Researchers; D42 ( b) Steinhart Aquarium/Tom McHugh/Photo Researchers; D47 Thomas Ives/The Stock Market; D48 (t) Erich Hartmann/Magnum Photos; D48 (b) Ron Sefton/Bruce Coleman, Inc.; D52-D53Ton Kinsbergen/ESA/Science Photo Library/Photo Researchers; D54 Tom Till; D57 Frank Zullo/Photo Researchers, Inc.; D58-D59 M. Agliolo/Photo Researchers, Inc.; D60 NASA; D63 (t) U.S. Geological Survey/Science Photo Library/Photo Researchers; D63 (b) David Crisp and the WFPC2 Science Team (Jet Propulsion Laboratory/California Institute of Technology); D63 (ct) NASA; D63 (cb) National Oceanic and Atmospheric Administration; D64 NASA; D64 NASA; D64-D65Erich Karkoschka (University of Arizona Lunar & Planetary Lab) and NASA; D65 (r) Dr. R. Albrecht, ESA/ESO Space Telescope European Coordinating Facility; NASA; D65 (c) Lawrence Sromovsky (University of Wisconsin - Madison), NASA; D66 NASA; D66 NASA; D66 NASA; D67 NASA; D68 NASA; D70 (r) Michael Freeman; D70 (tl) David Nunuk/Science Photo Library/Photo Researchers; D70 ((bl)) Omikron Collection/Photo Researchers; D71 (t) Simon Fraser/Science Photo Library/Photo Researchers; D71 (b) Robert Frerck/Tony Stone Images; D71 (ti) Roger Ressmeyer/Corbis; D72 NASA; D73 NASA; D73 NASA; D73 NASA; D74 NASA; D74 NASA; D74 NASA; D75 NASA; D76 Michael Holford; D78 John Sandford/Science Photo Library/Photo Researchers; D79 (t) Jerry Schad/Photo Researchers; D79 (b) John Sanford/Science Photo Library/Photo Researchers D82 (l) The Granger Collection, New York; D82 (r) Jean-Loup Charmet/Photo Library/Photo Researchers; D82-D83 (bg) NASA; D83 (l) Sylvester Allred/Visuals Unlimited; D83 (r) Mark E. Gibson/Dembinsky Photo Associates; D84 (t) J. Kelly Beatty/Sky Publishing Corporation; D84 (b) Science VU/Visuals Unlimited.

**Unit E**

E2-E3Jon Riley/Tony Stone Images; E3 (l) Dr. E.R. Degginger/Color-Pic; E3 (r) Dr. E.R. Degginger/Color-Pic; E4 Superstock; E6 Michael Denora/Liaison International; E8(b) Bob Abraham/The Stock Market; E10 Robert P. Carr/Bruce Coleman, Inc.; E14-E15(b) Richard R. Hansen/Photo Researchers; E16 Tony Stone Images; E20 (t) Kathy Ferguson/PhotoEdit; E20 (b) Doug Perrine/Innerspace Visions; E20 (bi) Felicia Martinez/PhotoEdit; E21 (b) Chip Clark; E22-E23Richard Pasley/Stock Boston; E24 Courtesy of J. G.'s Edible Plastic; E25 David R. Frazier; E26 (b) United States Nuclear Regulatory Commission; E26 (b) Tom Carroll/Phototake; E30-E31Ray Ellis/Photo Researchers; E31 (t) Peter Steiner/The Stock Market; E31 (b) Murray & Assoc./The Stock Market; E32 Craig Tuttle/The Stock Market; E35 (t) Jim Zipp/Photo Researchers; E36 Ted Horowitz/The Stock Market; E40 D. Nabokov/Gamma Liaison; E42 (b) L. West/Bruce Coleman, Inc.; E42 (i)Jonathan Wright/Bruce Coleman, Inc.; E43Gary Milburn/Tom Stack

& Associates; E44 (b) Jeff Foott/Bruce Coleman, Inc.; E48 (t) Craig Hammell/The Stock Market; E48 (b) Russell D. Curtis/Photo Researchers; E49 (tl) Stu Rosner/Stock, Boston; E49 (tr) John Mead/Science Photo Library/Photo Researchers; E49 (br) John Cancalosi/Stock, Boston; E50 (b) David Falconer & Associates; E50 (bi) Montes De Oca & Associates; E50 (tr) Telegraph Colour Library/FPG International; E50 (c) Charles D. Winters/Photo Researchers; E51 Paul Shambroom/Science Source/Photo Researchers; E53 Danny Daniels/The Picture Cube; E54 (t) Minnesota Historical Society; E54 (b) Peter Vadnai/The Stock Market; E58-E59Stephen Dalton/Photo Researchers; E59 Carl R. Sams, II/Peter Arnold, Inc.; E60 A. Ramey/PhotoEdit; E63 (li)) Michelle Bridwell/PhotoEdit; E63 (ri) Peter Langone/International Stock Photography; E63 (b) Summer Productions; E66 Randy Duchaine/The Stock Market; E68 (l) David Barnes/Tony Stone Images; E68 (r) Jim Zipp/Photo Researchers; E71 (b) Hank Morgan/Science Source/Photo Researchers; E72 Stocktrek/Frank Rossotto/The Stock Market; E74 Brian Parker/Tom Stack & Associates; E80 Bruce Forster/Tony Stone Images; E81 Russ Berger Design Group; E82 (b) Bose Corporation; E82 (b) Bose/Lisa Borman Associates; E86-E87Pete Saloutos/The Stock Market; E82 (t) Bose Corporation; E82 (b) Bose/Lisa Borman Associates; E86-E87 Pete Saloutos/The Stock Market; E88 Doug Martin/Photo Researchers; E93 Charles D. Winters/Photo Researchers; E94 Cosmo Condina/Tony Stone Images; E100 National Maritime Museum Picture Library; E103 (tr) Richard Megna/Fundamental Photographs; E103 ((bl)) Richard Megna/Fundamental Photographs; E103 (br) Richard Megna/Fundamental Photographs; E103 (cr) Richard Megna/Fundamental Photographs; E104-E105 (t) Phil Degginger/Color-Pic; E106 Gamma/Liaison International; E108-E109 Spencer Grant/PhotoEdit; E109 (t) Tom Pantages; E109 (br) Tom Pantages; E111 (t) Phil Degginger/Color Pic; E111 (c) Phil Degginger/Color Pic; E111 (b) Bruno Joachin/Liaison International; E111 (bg) W. Cody/Corbis Westlight; E114 (r) Corbis-Bettmann; E114 (l) Phil Degginger/Color-Pic; E114 (t) William E. Ferguson; E115 (b) Phil Degginger/Color-Pic; E116 (t) Fonar Corporation; E116 (b) Jean Miele/The Stock Market.

**Unit F**

F2-F3 PictureQuest; F3 (tc) Dwight R. Kuhn; F3 (br) Tony Freeman/PhotoEdit; F3 (li) Dwight R. Kuhn; F4 (bl)David R. Frazier; F6 (b) Mark E. Gibson; F8 (b)Bob Daemmrich/Stock Boston; F10 (bl) Miro Vintoniv/Stock Boston; F12 (bl) Jean-Marc Barey/Agence Vandystadt/PhotoResearchers; F13 (tr) Daniel MacDonald/The Stock Shop; F15 (tl) Bernard Asset/Agence Vandystadt/Photo Researchers; F15 (b) Bernard Asset/Agence Vandystadt/Photo Researchers; F16 (t) Kathi Lamm/Tony Stone Images; F20(b) William R. Sallaz/Duomo Photography; F23 (c)Photo Library International/ESA/Photo Researchers; F23 (cr) Photo Researchers; F26 (bl)) Michael Mauney/Tony Stone Images; F28 Brian Wilson; F29 (t) PA News; F29 (b)Michael Newman/PhotoEdit; F30 (t) UPI/Corbis-Bettmann; F34-F35 (bg) Dan Porges/Bruce Coleman, Inc.; F35 (tl) The Granger Collection; F35 (tr) R. Sheridan/Ancient Art and Architecture Collection; F36 (bl)) William McCoy/Rainbow; F38(bl)Yoav Levy/Phototake/PictureQuest; F41 (tr) Michael Newman/PhotoEdit; F42(tr) David R. Frazier; F46 (b) Mark E. Gibson; F48 (b) Jeff Dunn/Stock Boston; F50 (bl)) Tom King/Tom King, Inc.; F52 (b) Aaron Haupt/David R. Frazier; F53 (tl) Dan McCoy/Rainbow; F53 (cl) David Falconer/Folio; F53(br) Michael Newman/PhotoEdit; F54(t) Superstock; F55 (r) Churchill & Klehr; F55 (bl) Staircase & Millwork Corporation, Alpharetta, GA; F56(br) Tony Freeman/PhotoEdit; F58 (l) Archive Photos; F58 (bl) Alexandra Guest/John F. Coates; F58 (t) Noble Stock/International Stock Photography; F59(l) Archive Photos; F59 (br) Eric Sanford/International Stock Photography; F60(bl) Library of Congress.

**Health Handbook:** R23 Palm Beach Post; R27 (t) Andrew Spielman/Phototake; (c) Martha McBride/Unicorn Stock; (b) Larry West/FPG International; R28 (l) Ron Chapple/FPG; (c) Mark Scott/FPG; (b) David Lissy/Index Stock.

**All other photographs by Harcourt photographers listed below, © Harcourt:**
Weronica Ankarorn, Bartlett Digital Photograpy, Victoria Bowen, Eric Camden, Digital Imaging Group, Charles Hodges, Ken Karp, Ken Kinzie, Ed McDonald, Sheri O'Neal, Terry Sinclair

**Art Credits**

Mike Dammer A25, A57, A83, A109, B35, B63, B93, C27, C55, C77, D27, D49, D85, E27, E55, E83, E117, F31; Jean Calder A91, A92, A96 (t), A102, A103; Susan Carlson A49, D22; Daniel del Valle F7; John Downes F56 (b); John Francis B14; Lisa Frasier E48-E49, D10; George Fryer C6, C18, C19, C42, C44, B70, D20, D21, D80, F46; Geosystems A49, B20; Patrick Gnan F42, E52, E3, E87; Pedro Julio Gonzalez A14; Terry Hadler E14, E19, E41, F46, F47; Tim Hayward A8, C70, C72; Robert Hynes B26, B28, B56, Joe LeMoniier A2, A56, E59, A82; Sebastian Quigley B86, E6, E8, E22, E90, E91, E104, E108, D12, D44, D62, D63, D64, D72, F22, F24, F25; Mike Saunders A18, A19, A65, A76, C7, C48, C62, C68, B43, B44, D8, D56, D57, D58; Steve Seymour C8, C13, B7, B80, E35, E36, E43, E62, E68, E69, E70, E75, E78, E79, E92, E96, E97, D14, D15, D43, D72, F15, F41; Eberhand Reinmann A90,A92, A96 (b), A98, A104, E64; Steve Westin C12, C20, C21, C40, C43, B48, E76, E77, E102, F48, F54, F55, F56 (t)